The Mosquito Crusades

Studies in Modern Science, Technology, and the Environment

Edited by Mark A. Largent

The increasing importance of science and technology over the past 150 years—and with it the increasing social, political, and economic authority vested in scientists and engineers—established both scientific research and technological innovations as vital components of modern culture. Studies in Modern Science, Technology, and the Environment is a collection of books that focuses on humanistic and social science inquiries into the social and political implications of science and technology and their impacts on communities, environments, and cultural movements worldwide.

Mark R. Finlay, *Growing American Rubber: Strategic Plants and the Politics of National Security*

Gordon Patterson, *The Mosquito Crusades: A History of the American Anti-Mosquito Movement from the Reed Commission to the First Earth Day*

The Mosquito Crusades

A History of the American Anti-Mosquito Movement from the Reed Commission to the First Earth Day

GORDON PATTERSON

RUTGERS UNIVERSITY PRESS
New Brunswick, New Jersey, and London

Library of Congress Cataloging-in-Publication Data

Patterson, Gordon M.
The mosquito crusades : a history of the American anti-mosquito movement from the Reed Commission to the first Earth Day / Gordon Patterson.
p. ; cm. — (Studies in modern science, technology, and the environment)
"Sequel to The Mosquito Wars: A History of Mosquito Control in Florida, which appeared in 2004"—Pref.
Includes bibliographical references and index.
ISBN 978–0-8135–4534–9 (hardcover : alk. paper)
1. Mosquitoes—Control—United States—History—20th century. 2. Mosquitoes as carriers of disease—United States. I. Patterson, Gordon M. Mosquito wars. II. Title. III. Series.
[DNLM: 1. Mosquito Control—history—United States. 2. History, 20th Century—United States. QX 11 AA1 P317m 2009]
RA640.P25 2009
614.4'323—dc22 2008038705

A British Cataloging-in-Publication record for this book is available from the British Library

Visit our Web site: http://rutgerspress.rutgers.edu

Manufactured in the United States of America

For Ben

Act well your part, there all the honour lies.

ALEXANDER POPE

To forget one's purpose is the commonest form of stupidity.

FRIEDRICH NIETZSCHE

Contents

Illustrations

Acknowledgments

THIS BOOK IS A SEQUEL to *The Mosquito Wars: A History of Mosquito Control in Florida*, which appeared in 2004. I thought I would be through with mosquitoes at that point. Instead I discovered that my interest in mosquito control had both deepened and broadened. Readers of *The Mosquito Wars* asked provocative questions about the development of mosquito control in other parts of the country. If mosquito control was such a good idea, one reviewer wrote, why did the mosquito warriors have to work so hard to win the public's assent? This question led to other questions. When and where did the American mosquito crusade begin? How did something that was an essentially local issue grow into a national movement? The Mosquito Crusades represents my effort to answer these questions.

Some explanation for my use of the word "crusade" is necessary. Since 9/11 the word "crusade" has fallen into disfavor. At the beginning of the twentieth century, progressive reformers called for temperance, women's suffrage, reform of election laws, trust busting, and, yes, a mosquito crusade. Until the late 1950s the mosquito control movement described itself as a crusade. In 1901 Leland Howard, then chief entomologist for the United States Department of Agriculture (USDA) called for an "intelligent crusade" against mosquitoes. Decades later, Howard (who lived to be nearly one hundred years old) maintained that no movement held greater consequence for Americans' health and welfare than the "mosquito crusade." My objective is to provide a perspective on the movement's "intelligence" as well as the historical consequence of the "crusade."

I am indebted to the entomologists, mosquito control workers, public health personnel, state officials, representatives of federal agencies, librarians, archivists, engineers, biologists, environmental activists, and historians who assisted me in my research. I wish to thank Dean Robert Goodman at Rutgers for allowing me to examine John B. Smith's papers and the historical archives located at the Headlee Laboratory and Blake Hall in New Brunswick.

This book would not have been possible without the support and access to the archives of the American Mosquito Control Association, the New Jersey Mosquito Control Association, the Utah Mosquito Abatement Association, the Florida Mosquito Control Association, the Louisiana Mosquito Control Association, the Mosquito and Vector Control Association of California, and the Robert Washino Library at the Sacramento–Yolo County Vector Control in Elk Grove, California. Versions of some the material presented in this book were delivered as lectures and conference papers and appeared in *Wing Beats* and conference proceedings. The discussion of Florida was drawn from my 2004 *Mosquito Wars: A History of Mosquito Control in Florida*, with the permission of the University Press of Florida.

I owe special thanks to Doug Carlson, the director of the Indian River Mosquito Control District, for his encouragement. I would be remiss if I did not acknowledge the help of Ryan Arkoudas, Richard Baker, John Beidler, Patti Bower, David Brown, Gil Challet, Gary Clark, Glen Collett, Joe Conlon, Roxanne Rutledge Connelly, Scott Crans, Wayne Crans, Major Dhillon, Sammie Dickson, Bruce Eldridge, Howard Emerson, Kellie Etherson, Ary Farajollahi, Randy Gaugler, Judy Hansen, Gary Hatch, Edward Kalajian, Mary Beth Kenkel, Uriel Kitron, Minoo Maden, Janet McAllister, William Meredith, Ken Minson, Chet Moore, Charles Myers, Roger Nasci, Lewis Nielsen, William Opp, Tadhgh Rainey, William Riesen, Henry Rupp, John Rusmisel, Joseph Sanzone, Mike Shaw, Donald Sutherland, Walter Tabachnick, Robert Taylor, Doreen Valentine, Robert Washino, and Matt Yates. I am grateful to my mother, Molly Patterson, and my wife, Joy Patterson, for their encouragement and wise counsel.

The Mosquito Crusades

Introduction

The Guardians of Paradise

At 1:00 p.m. EST on November 10, 1951, Leslie Denning, mayor of Englewood, New Jersey, phoned Frank Osborn, the mayor of Alameda, California. In Englewood, nearly one hundred reporters had crowded into the mayor's office to watch Denning make the nation's first long-distance, direct-dial phone call. Three thousand miles to the west in Alameda, Osborn waited expectantly. When the phone rang, Osborn answered declaring, "This is a great thing for both our cities. New Jersey to Alameda in eighteen seconds. The world shrinks so that soon there won't be enough room for the people." During the next few minutes smiling members of both the Englewood and Alameda chambers of commerce listened as the mayors extolled their hometowns' virtues and traded quips over the quality of life on the east and west coasts. At the phone call's conclusion Osborn fired a parting tease, "Is it true that people in New Jersey ride mosquitoes the same as we ride horses out here?"[1]

None of the newspaper accounts of the historic phone call included Denning's response. Months later Mrs. Harvey Bein, chairperson of the New Jersey Federation of Women's Clubs committee on mosquito control, sought to set the record straight. "The deadly effect of [the mosquito's] bite on the health and welfare of the people," Mrs. Bein told the participants at the thirty-ninth annual meeting of the New Jersey Mosquito Extermination Association, "does not provoke one who knows of it to puns and laughter. You will be interested to know that while the humor has not been entirely dissipated, your good work is being spread to the other side of the continent." Denning had not allowed Osborn's taunt, Mrs. Bein reported, to pass unchallenged. "I haven't seen any [mosquitoes] around here lately," Englewood's mayor had shouted in an attempt to make himself heard over the reporters' guffaws.[2]

To a generation accustomed to instant messaging, cell phones, and video conferencing, the excitement about a direct-dial, long-distance phone call may seem a quaint reminder of a less complicated technological era. Few would dispute the link between the first direct-dial phone call and the ubiquitous wireless networks that define the contemporary world.

But the reference to mosquitoes is puzzling. Most Americans in recent decades give little thought to mosquitoes. This was not the case throughout most of American history. The first European travelers to the New World commented on the prodigious numbers and ferocity of America's indigenous mosquitoes. In the seventeenth century the early English colonies at Jamestown and Plymouth suffered greatly from them. George Washington and Abraham Lincoln grumbled about mosquitoes. In 1879 mosquitoes forced Rutherford Hayes to retreat from the nation's capital to Fremont, Ohio.[3] A generation earlier a Virginia congressman opposed Florida's entry into the Union because of mosquitoes. Nothing wholesome, John Randolph opined, would come from Florida because it was "a land of swamps, of quagmires, of frogs, and alligators, and mosquitoes."[4]

Throughout the country's history, most Americans assumed that mosquitoes were like the weather. Everyone complained, but nothing could be done to relieve the mosquito blight. At the beginning of the twentieth century, a handful of progressively minded individuals in New Jersey declared their intention to do something about mosquitoes. Led by a self-trained entomologist named John B. Smith, they opened a campaign against the mosquito pest. Their efforts grew into a nationwide crusade.

No country in history has waged so vast a war against mosquitoes as the United States. By 1910 a mosquito crusade had spread from the Atlantic to the Pacific coasts. Initially the campaign grew out of a desire for relief from pest mosquitoes. Support for mosquito control dramatically increased after the discovery of the role that mosquitoes play in spreading diseases such as malaria and yellow fever.

The word "mosquito" bears a distinctly American accent. At least, this is Samuel Rickard Christophers's argument in his introduction to his monumental examination of the life history and bionomics of the *Aedes aegypti* mosquito. "With occasional exceptions," Christophers maintains, it was possible until the end of the nineteenth century to determine whether a speaker was English or American by their "respective use of gnat or mosquito."[5] British entomologists employed the word "gnat" for the double-winged, biting flies that periodically rose above the English fens. They reserved the word "mosquito" for the larger, more ferocious, oversexed pest described in travelers' tales from the New World.

The word "mosquito," in fact, was first employed in the Americas.[6] Lexicographers trace its origin to sixteenth-century Spanish and Portuguese mariners who combined the Spanish "mosca" (fly) with the diminutive suffix "ito" (little). In 1535 Fernandez de Oviedo y Valdes, commander of the Spanish garrison at Santo Domingo on the island of Hispaniola, provided the first detailed description of the American "musketa." Oviedo wondered

> how such a tiny creature can have reason or strength, their perfection is so inexplicable and incomprehensible, for it is said of the mosquito, Where could Nature place so much intelligence in the mosquito, the 'Zunsal' (the one that sings)? Where did she put their sight? Where their taste? Where their sense of smell? Where did she engender a voice so awful in comparison with the small size of its body? With what subtlety she fastened on her wings and formed those long legs, and the belly hungry and greedy for human blood, or with what artifice did she sharpen its needle, that even though it is so fine as to be invisible it is capable of piercing the skin, and is hollow for sucking blood.[7]

Fifty years elapsed before the first published reference to mosquitoes appeared in English. In 1589 Miles Phillips complained in *Hakluyt's Voyages* "we were almost oftentimes annoyed with a kind of flie [fly], the Spaniards call musketas [mosquitoes]."[8]

French, German, and Swedish colonists invented words to describe the "little demons" that tormented their efforts to subdue the New World. Not satisfied with a single word, French explorers and settlers alternately referred to mosquitoes as "les cousins," "les moucherons," and "les maringouins."[9] Germans in Pennsylvania railed against the "Stechmucken" that plagued their days and nights. In 1643 the Swedes erected Fort Elfsborg on the Delaware River at Elsinboro Point to harass Dutch trading vessels. "Mygge" (Swedish for "mosquitoes") were so numerous that the fort was soon nicknamed "Myggenborg."[10]

If the Europeans could not agree on a name for the pest, they exhibited a remarkable consistency in their portrayals of the blood sucking pests' fury. Dutch settlers in what is now Bergen County, New Jersey, spoke of mosquitoes as large as sparrows.[11] Farmers along the New England and mid-Atlantic coasts complained that when they came down to the salt marshes to cut summer hay, the meadows charged one mosquito bite for each spear of cut grass.[12] The song of the New World's mosquitoes put their European cousins to shame. "The noise they make in flying," English observers reported, "cannot be conceived by persons who have only heard gnats in England."[13]

Catholic priests sent to evangelize the new continent left numerous harrowing accounts of their battles with mosquitoes. In September 1727 a French Jesuit named DuPoisson, traveling down the Mississippi River, described mosquitoes as penance for the sins of the world. "The greatest torment in comparison with which all the rest would be but sport," Father DuPoisson declared, "is the mosquitoes—the cruel persecution of the mosquitoes. The plague of Egypt I think was no more cruel. . . . This little insect has caused more swearing since the French have been in Mississippi, than had previously taken place in all the world."[14]

Ninety years later the German naturalist Alexander von Humboldt had difficulty in finding words to describe the situation of Spanish priests on the Orinoco River. "Persons who have not sailed on the great rivers of equinoctial America," Humboldt observed, "cannot conceive that at every moment without interruption, one can be tormented by the insects swarming in the land and how the numbers of these little animals can render vast regions almost uninhabitable." Later in his journey, Humboldt came on an "old Spanish missionary who told us with an air of sadness that he had passed *his twenty years of mosquitoes* in America [emphasis in original]. He told us to look at his legs in order that we might be able to tell the people some day across the sea what the poor monks suffer in the forests of the Cassiquiare."[15]

Protestant English colonists in temperate North America were not exempted from the persecution of mosquitoes. The Virginia Company's fortune seekers at Jamestown and Puritan divines at Plymouth complained mightily of them. On December 20, 1606, 120 adventurers set sail from London for Virginia aboard the three tiny ships. After exploring the Chesapeake capes, they made what historian Samuel Eliot Morison describes "as the usual mistake of firstcomers to America by settling on a low, swampy island."[16]

It was a miserable choice. A popular contemporary ballad had proclaimed "Virginia, Earth's only Paradise," yet hordes of bloodthirsty mosquitoes rose from the Chesapeake marshes to greet the English adventurers. In the first six months after Jamestown's founding, 51 of the colonists perished from "fevers." Two-thirds of the 1,700 migrants who came to Jamestown perished in the next decade from the periodic outbreaks of a "devastating fever that raged in Jamestown in the early years of its settlement."[17]

Conditions were no better in Massachusetts. William Bradford, the thirty-one-year-old governor of the Plymouth Colony, noted with displeasure his comrade's complaints about mosquitoes. Some grumbled that Plymouth bore a closer resemblance to the pharaoh's court that had been "corrupted by reason of the swarm of flies" (Exodus 8:24) than to the promised New Zion. On January 24, 1623, Bradford penned an exasperated note in his *History of the*

Plymouth Colony. "The question which has brought up against the Plymouth Colony," he observed, "was that the people are much annoyed by mosquitoes. The answer: They are too delicate and unfit to begin new plantations and colonies that cannot endure the biting of the mosquito. We wish such to keep at home until at least they be mosquito proof."[18]

Eight years after Bradford's complaint, a twenty-seven-year-old Puritan divine named John Eliot arrived in Roxbury, Massachusetts, and set about reforming the native Algonquins' heathen anti-mosquito ways. By 1647 Eliot had mastered sufficient Algonquin to preach in the nearby Native American villages. Before baptism, Eliot required that his Algonquin converts "must wear their hair English-style, stop using bear's grease as protection against mosquitoes, and give up plural wives."[19] Historian Colin Calloway noted in his study of early contact between Europeans and Native Americans that Eliot's effort to convince the Indians to abandon plural marriage and their anti-mosquito repellents met with strong resistance. Reportedly Eliot rebutted these objections by declaring, "prayer and pains through faith in Christ Jesus will accomplish anything."[20]

Missionaries, fur trappers, and settlers recorded numerous accounts of the Native Americans' exotic strategies to reduce the mosquito blight. In 1699 the Quaker Jonathan Dickinson and his family were shipwrecked near Jupiter Inlet on Florida's Atlantic Coast. Dickinson reported that the torment from swarms of salt marsh mosquitoes grew so intense that he and his companions adopted the local Indian strategy and "digged holes in the sand, got some grass and laid it therein to lie on, in order to cover ourselves."[21] A generation later William Byrd II, a member of the King's Council and founder of Richmond, Virginia, confirmed Eliot's report that Native Americans relied on bear's grease "as a general defence against every species of vermin. . . . They say it keeps both bugs and mosquitoes from assaulting their persons."[22]

Reports of Native Americans' use of unguents and smoke to drive off mosquitoes continued well into the nineteenth century. William Dewees, who supported Stephen Austin's effort to colonize Texas, observed that the native population "went about naked but painted their faces, tattooed and pierced their bodies, and smeared alligator grease on themselves to ward off mosquitoes."[23] Settlers reported that the Timucua, who lived along the Atlantic Coast in south Georgia and north Florida, regularly placed "smudge pots under their beds" to prevent mosquitoes biting during the night.[24]

Fabulous reports of mosquitoes' ferocity persisted throughout the eighteenth and nineteenth centuries. Isaac Weld, writing of upstate New York in his *Travels through North America: 1795–1797*, declared, "General Washington *told me* [emphasis in original] that he never was so much annoyed by mosquitoes in any part of America as in Skenesborough, for that they used to bite through the thickest boot." A century later, Leland Howard, the federal government's chief entomologist, commented on this passage that "knowing boots of those days were very thick, and that the mosquitoes of that time must have been structurally identical with those of today, there arises instantly the question of the veracity between Mr. Weld and General Washington." Howard concluded that because "we know from Dr. Weem's veracious history that General Washington was so constituted that he could not tell a lie, it looks very much as though Mr. Weld, like certain other travelers who have written books on their return home after extended travel, was somewhat inclined to overstate the truth."[25]

Washington was not alone among American presidents to call attention to the country's most prominent pest. Abraham Lincoln boasted of being a dedicated mosquito warrior. In the spring of 1832, a twenty-three-year-old Lincoln served in the Illinois militia during the three-week-long Black Hawk War. Lincoln did not see a single hostile Native American while on active duty. Nonetheless, he later told potential voters "I fought, bled, and came away. I had," he explained, "many bloody struggles with mosquitoes; and although I never fainted from loss of blood, I can truly say I was often hungry."[26]

At the beginning of the twentieth century, most Americans considered mosquitoes, like death and taxes, an unavoidable component in the human condition. At Kitty Hawk, North Carolina, only unfavorable weather and mosquitoes could keep the Wright brothers from working. Orville Wright wrote plaintively to his sister Katherine in 1900, describing North Carolina's winged pests. "A swarm of mosquitoes came in a mighty cloud, almost darkening the sun. This was the beginning of the most terrible experience I ever passed through. They chewed clear through our underwear and socks. Lumps began swelling up all over my body like hen's eggs."[27]

Wilbur and Orville Wright returned to the Outer Banks in the summer of 1901. They came at a bad time. It had rained for most of the previous three weeks and the mosquitoes started to hatch out. The Wright brothers persevered. Two years later on December 17, 1903, Orville Wright made his historic twelve-second flight. Neither of the Wright brothers knew that a handful of equally intrepid individuals had already launched a campaign to rid America of the mosquito pest.[28]

The formal science of medical entomology originated in the 1870s and 1880s. In 1877–1878 Patrick Manson, a Scottish physician serving as a medical officer for the Chinese Imperial Maritime Customs Department in Formosa, identified the nematode worms responsible for filariasis in a mosquito's gut. "I shall not easily forget the first mosquito I dissected," Manson reminisced.

"I tore off its abdomen and succeeded in expressing the blood the stomach contained. Placing this under the microscope, I was gratified to find that, so far from killing the *filaria*, the digestive juices of the mosquito seemed to have stimulated it to fresh activity."[29]

Manson's experiments opened a new epoch in the study of tropical medicine.[30] Three years later in 1881 a Cuban physician named Carlos Finlay shocked his colleagues when he declared: "I feel convinced that any theory which attributes the origin and propagation of yellow fever to atmospheric or meteorological conditions (miasma), to filth, or to the neglect of general hygienic precautions, must be considered as utterly indefensible." Finlay contended that yellow fever was transmitted by "something tangible," the *Culex* mosquito (*Aedes aegypti*).[31] He was aware that there was powerful resistance to his hypothesis. "I understand but too well," he declared in an 1881 paper, "that nothing less than an absolutely incontrovertible demonstration will be required before the generality of my colleagues accept a theory so entirely at variance with the ideas which have until now prevailed about yellow fever."[32] Despite repeated experiments during the next nineteen years, Finlay failed to provide conclusive scientific proof for his hypothesis.

Two years after Finlay published his claim that mosquitoes transmitted yellow fever, a Washington, D.C., obstetrician named Albert Freeman Africanus King delivered a talk in which he listed nineteen reasons why mosquitoes were responsible for spreading of malaria. King, best remembered for treating Lincoln the night he was assassinated, developed his mosquito theory in response to a call from the U.S. Sanitary Commission on how to respond to the growing malaria problem in the nation's capital.[33]

In 1882 King presented before the Washington Philosophical Society the first of two papers arguing that mosquitoes transmitted malaria. He had assembled a mass of evidence demonstrating that malaria was "produced by a poison injected into the human system by mosquitoes." Unfortunately, King took his argument too far. He contended that this "poison . . . in the absence of quinine, affects the spleen in such a way as to induce the formation of the pigment which colors the negro's skin." King believed that "the whiteness of the white American is due wholly to quinine and that if nature and mosquitoes could have fair play he would be as black as the blackest of plantation negroes."[34] To protect the white citizens of the nation's capital, King recommended that Congress erect a "colossal woven-wire screen as high as the Washington Monument" around the capital. Paul Russell thought that King's observations "cast doubt on his sanity."[35]

Medical entomologists credit Ronald Ross, an English physician serving in the India Medical Service, with unraveling the malaria riddle. While on home leave from India in 1894, Ross visited Patrick Manson's laboratory in London. Manson showed Ross the tiny filarial worms and explained that "mosquitoes were simply nursemaids to the filariae."[36] Manson encouraged Ross to pursue his research on the role of mosquitoes in the propagation of malaria when he returned to Secunderabad. On August 20, 1897 (henceforth, Mosquito Day), Ross made his crucial discovery. In July 1898 Ross wrote to Manson that he felt "*almost* [emphasis in original] justified in saying that I have completed the life cycle, or rather perhaps one life-cycle" of the malaria parasite.[37]

While Ross, in India, and Battista Grassi, an Italian zoologist who simultaneously discovered the malaria plasmodium in the digestive tract of *Anopheles* mosquitoes, were sorting out the final strands of the human-mosquito-human malaria cycle, Americans launched a war against Spain.[38] More than 270,000 Americans served in the 114-day Spanish American War. Only 379 of the 5,462 Americans casualties died in combat. Malaria, typhoid, dysentery, or yellow fever caused the overwhelming majority of American causalities in the nation's "splendid little war" with Spain.[39]

When the fighting stopped, a young army physician named Walter Reed decided to launch a "thorough investigation of Yellow Fever in Cuba." Reed feared that yellow fever and malaria would undermine the U.S. victory. Ross and Grassi's findings had sparked Reed's interest in the role of mosquitoes in transmitting diseases. Two years later, Reed's colleague Dr. Jesse Lazear visited Leland Howard at the United States Department of Agriculture (USDA) to learn more about malarial mosquitoes. Howard had recently published an article in *Scientific American* describing the life history of the *Anopheles quadrimaculatus* mosquito, one of the major American vectors of malaria.[40]

Howard's article and his lectures on mosquitoes generated considerable interest in the role of mosquitoes in spreading disease. "It resulted from all of this," Howard later explained in his autobiography *Fighting Insects: The Story of an Entomologist*, "that Dr. Walter Reed . . . came to my office to make a study of certain mosquitoes prevalent in Cuba."[41] In the summer of 1900 the U.S. Army's surgeon general, George Sternberg, asked Reed to lead the Army's investigation of yellow fever in Cuba. Reed arrived in Cuba in June 1900. During the next six months, the Reed Commission wrestled with the problem of yellow fever. Howard reported that Reed "constantly wrote to me, not only on questions relating to the different species of mosquitoes, but also to tell me of the progress of [the Commission's] work."[42]

Reed pressed Howard for information about the *Culex fasciatus* mosquito [now *Ae. aegypti*]. Once in Cuba, Reed, Lazaer, Charles Carroll, and Aristides Agramonte met with Finlay. Finlay volunteered his services to the

Commission and encouraged them to test his hypothesis that *Culex fasciatus* was the vector for yellow fever. In January 1901 Reed wrote to Howard "of course, you have already heard from Dr. Sternberg of our complete success. . . . The mosquito theory for the propagation of yellow fever is no longer a *theory* but a *well established fact* [emphasis in original]. Isn't it enough to make a fellow feel happy? *Anopheles* and *Culex* [*Ae. aegypti*] are a gay old pair! What havoc they have wrought on our species during the last three centuries! But with "*Howard* and *kerosene* [emphasis in original], we are going to knock them out!"[43]

The Reed Commission's report and Howard's recommendations guided William Gorgas, an Army physician, in his effort to eliminate yellow fever in Havana. By 1902 Gorgas's anti-mosquito campaign had succeeded in eradicating the threat of yellow fever and in substantially reducing the malaria rate. Two years later General Sternberg charged Gorgas with the task of developing an anti-mosquito campaign to protect the workers in the Panama Canal Zone.

At the beginning of the twentieth century Leland Howard articulated an entomological precept that would echo through the next seventy years: "The insect world is a menace to the dominance of man on this planet." To Howard and his successors the struggle of human beings against insects in general and mosquitoes in particular was no less than a war "which will in all probability end in the elimination of one side or the other."[44]

The announcement of the findings of the Reed Commission and the success of William Gorgas in eliminating yellow fever and malaria in Havana and Panama encouraged a group of entomologists, engineers, and progressive reformers in New Jersey to launch a nationwide anti-mosquito crusade. The Reed Commission gave the crusade credibility and a goal. "The perceived success of the campaigns in Cuba and Panama," Linda Nash observed in her ecological history of California's Central Valley, "were instrumental in elevating the strategy of insect elimination to the position of gospel in American public health."[45]

Like the Christian Gospels, the gospel of the anti-mosquito movement was open to interpretation. Between 1900 and 1970 disagreements emerged among the crusaders over the movement's techniques, organization, and objectives. On one level, local health boards and many physicians argued that anti-mosquito work should target only those species of mosquitoes that were competent disease vectors. Others advocated comprehensive control strategies maintaining that all species of mosquitoes posed a potential health risk.

The engineers, physicians, entomologists, business executives, and club women who led the mosquito crusade were part of the Progressive movement advocating government initiatives to eliminate graft and corruption, regulate the railroads, and improve the quality of life. These individuals shared two primary motivations for their work. Historically the anti-mosquito movement grew out a desire of white, middle-class professionals who were primarily concerned with the reduction of pest and nuisance mosquitoes. These individuals believed that effective mosquito control would increase the value of real estate, encourage tourism, and stimulate economic growth. Later, public health workers employed a similar economic argument to win the support of textile and mill companies for their anti-malaria work.

In 1900 John B. Smith, a Rutgers entomology professor, launched what was to become a nationwide anti-mosquito movement. Smith's research along the Jersey Shore convinced him that drainage offered a viable permanent solution to the Garden State's salt marsh mosquito problem. When Smith died in 1912 his work had encouraged anti-mosquito crusaders in California, Louisiana, and Florida.

The anti-mosquito movement evolved across two distinct phases: the drainage era and the insecticide era. During the era of ditches and drains, which lasted roughly from 1900 to the start of World War II, local mosquito commissions formed in New Jersey, California, Florida, Utah, and Illinois.

The era of insecticides began in 1942 at a laboratory in Orlando, Florida, when a group of USDA researchers discovered the toxicant properties of dichlorodiphenyltrichloroethane (commonly known as DDT) in killing both larval and adult mosquitoes.[46] The insecticide offered a way to protect American soldiers as they trained at home and, later, when they fought in Africa, Europe, and the Pacific.

The publication of Rachel Carson's *Silent Spring* in 1962 marked the beginning of the modern environmental movement and the end of the mosquito crusade. During the insecticide era, the anti-mosquito movement had expanded tremendously. The availability of powerful, inexpensive chemical means of control encouraged many established mosquito programs to abandon drainage and ditching work. The reliance on insecticides generated both biological and political resistance. Biologically, insects adapted to the toxicants while other species seemed to suffer their deleterious side effects. *Silent Spring* became a rallying point for the political resistance, opening a new chapter in human beings' relationship with the insect world. By 1970 environmentalists denounced the mosquito crusaders for their success at what earlier generations would have considered a beneficent act.

The history of the American mosquito crusade is complex and nuanced. This book chronicles the organization and development of mosquito control between 1900 and 1970. Whenever possible, I have sought to allow the individuals who dedicated themselves to waging war on mosquitoes and their critics to speak for themselves. The struggle between our species and the taxonomic family of mosquitoes has had profound effects on both human beings and our environment. What emerges is a story that is sometimes tragic, occasionally ridiculous, and often heroic.

Chapter 1 Waging War on the Insect Menace

ON MAY 16, 1901, a man named Spencer Miller drove his horse and buggy to the South Orange, New Jersey, train station to meet Leland Howard. Miller had invited Howard, the head of the Bureau of Entomology in the United States Department of Agriculture (USDA), to address a handful of progressive businessmen and civic leaders who sought practical advice on fighting the mosquito menace. South Orange was one of a string of villages and towns located along the first and second ridges of the Watchung or Orange Mountains. The handsome countryside and proximity to Newark, Elizabeth, and New York City led to rapid growth in the Oranges at the turn of the century. "There is a beauty," a contemporary observer noted, "in all these places and in many others not far off; but the blight of the mosquito pest lies over-all."[1]

The mosquito blight reached a decisive point in the 1890s. Property values plummeted and chambers of commerce feared that mosquitoes would stifle future growth. "Magnificent dwellings find no purchasers because of the necessity of close screening to keep out mosquitoes and because, as soon as dark sets in, piazzas must be abandoned to escape the annoyance of these little nuisances. In almost any car in any morning train to the city during the summer, somebody may be heard talking of mosquitoes, and when they are 'bad' one hears little else."[2]

In 1900 Spencer Miller, a successful engineer later famous for the invention of a cable carrier system employed in the construction of the Panama Canal and Hoover Dam, decided to build a house in the Montrose Park district of South Orange.[3] Work began in April. However, mosquitoes "so tormented the men" that Miller's immigrant carpenters threatened to quit. "I was foolish enough at the time," Miller recalled, "to say that there is an answer to this problem and mosquitoes can be gotten rid of, and of that I have received a good deal of ridicule."[4] In the following months Miller improvised numerous ineffectual anti-mosquito devices. His most imaginative invention employed a bright light to lure mosquitoes into a giant exhaust fan that promptly ground the six-legged beasts to oblivion. Thirty years later Thomas Headlee and Tommy Mulhern capitalized on mosquitoes' attraction to light with what became known as the New Jersey Light Trap. This light trap became the standard mosquito collection device for much of the twentieth century. Miller, though, had wanted to kill mosquitoes, not count them. None of his experiments proved satisfactory.

Shortly after moving into his South Orange home, Miller was invited by the engineering faculty at Cornell University to describe an innovative method of coaling naval vessels at sea that he had developed during the Spanish-American War. While in Ithaca, Miller learned that Leland Howard had recently delivered a spellbinding talk before the Cornell alumni association titled "Mosquitoes and Mosquito Extermination" drawn from his new book, which explained how mosquitoes could be controlled. When Miller returned to South Orange, he persuaded his business associates and professional friends to join him in inviting Howard to speak. They agreed to rent space in a hall and pay for Howard's trip from Washington to New Jersey. Miller and his comrades planned to use Howard's speech on May 16 as the clarion call for a mosquito crusade along the Jersey Coast.

Born in 1857 in Rockford, Illinois, Howard's career as an entomologist and public servant spanned more than a half century. In 1878 he joined the USDA. Sixteen years later he took charge of the USDA's entomological service. Howard published his pioneering *Mosquitoes: How They Live; How They Carry Disease; How They Are Classified; How They May Be Destroyed* in 1901. A decade later Howard, Harrison Dyar, and Frederick Knab published the first volume of their classic *Mosquitoes of North and Central America and the West Indies* (1912). In 1927 federal age restrictions forced Howard to step down as administrative head of the USDA's Bureau of Entomology although he remained the bureau's principal entomologist until his retirement in 1931.

If it had not been for his father's sudden death in 1873, Howard would almost certainly have followed his father's wishes and pursued a career in law. Instead the sixteen-year-old declared civil engineering as his major at Cornell. Difficulty with freshman calculus, however, led him to change his major to natural history.[5] Childhood books from his parents on butterfly collecting and harmful insects had kindled what was to become his life-long passion for entomology.[6] He graduated from Cornell in 1877. His mother despaired when he announced his intention to pursue a career in entomology. Family friends issued dire warnings. "They all said," Howard later wrote, "that I would never be rich and that they doubted that I would ever be able to marry and live comfortably."[7]

Howard launched his career as an entomologist in this context. In 1878 Charles Valentine Riley, Missouri's state entomologist, succeeded Townsend Glover as head of the newly formed entomological service within the USDA.[8] Riley asked John Comstock, Howard's professor at Cornell, to accept a temporary assignment as the entomological service's field agent. On his way south, Comstock stopped in Washington for Riley's final instructions. A new position had recently opened at the USDA, and Riley inquired if Comstock "had at Cornell an advanced student of good appearance, fairly good manners, and ability, who would come to Washington as his assistant."[9] Comstock recommended Howard. Riley wrote and offered the budding entomologist the position at a salary of one hundred dollars a month.

Howard's interest in mosquitoes intensified in the early 1890s. In his autobiography, Howard recounted his childhood fascination with mosquitoes and his initial experiments in 1867 using kerosene to kill mosquito larvae. Twenty-five years later Howard launched the USDA's first large-scale experiment testing the effectiveness of oil as a larvicide. In 1892 he published articles in *Insect Life* describing his experiments to determine the quantity, grade, and best method of applying oil as a larvicide. Entomologists praised his precision. Newspapers reported that Howard's experiments demonstrated that there was a remedy for the mosquito blight. An editorial writer for the *American Naturalist* applauded Howard's suggestions and hailed him for his discovery of oil's effectiveness in controlling mosquitoes.[10]

Howard never claimed this. In the *Insect Life* articles and later publications, he stressed that many other individuals had commented on oil's effectiveness in killing mosquitoes. In 1793, for example, in the midst of a devastating yellow fever epidemic that forced President George Washington and legislators to retreat to New York, a correspondent for the Philadelphia *Daily American Advertiser* observed "that a half teaspoon of any common oil" when poured on a rain barrel brought relief from the stinging pests.[11] So too did the Romantic poets Robert Southey and Samuel Taylor Coleridge propose "to diminish their [mosquitoes] numbers by pouring oil upon the great standing waters and large rivers, in those places most infested by them" in England's Lake District.[12]

When Howard became the USDA's chief entomologist, he expanded the entomological service's research on mosquitoes. Howard published in 1900 the USDA's first bulletin devoted exclusively to mosquitoes and mosquito control. In *Notes on the Mosquitoes of the United States* he reported that his office had received numerous letters from individuals who had employed oil as a larvicide. H. E. Weed told Howard how he had successfully used oil "to rid the college campus of Mississippi Agricultural College of mosquitoes by the treatment of eleven large water tanks with oil." On Long Island John B. Smith, Howard's former aide at the USDA, described two experiments using oil to control mosquitoes.[13]

Neither Howard nor any of his correspondents considered the potential negative environmental effects of using oil as a larvicide. There is evidence, however, that he did not consider oil a panacea. Oil, he advised, was best employed in certain specific locations such as sewers and "comparatively small swamps and circumscribed pools. It is not," he declared, "the great sea marshes along the coast, where mosquitoes breed in countless numbers, which we can expect to treat in this method."[14]

In 1901 Howard published his popular introduction to mosquito control. Written in a simple, direct style, Howard's descriptions of the twenty-five known species of American mosquitoes and their control was destined to become the "bible of all interested in mosquito warfare."[15] The book found an immediate audience. In Washington the Army's surgeon general, George Sternberg, ordered that copies of Howard's book be issued to all members of the Army Medical Corps. William Gorgas, who led the army's campaign against mosquito-borne diseases in Havana and later in Panama, praised the effectiveness of Howard's anti-mosquito recommendations.

Howard's publication of *Mosquitoes* prompted Spencer Miller's invitation. The South Orange meeting hall was filled when Miller and Howard arrived. After a brief introduction, Miller went to the back of the room to run the lanternslides that Howard had brought to illustrate his talk. Many newspaper reporters attended, Miller recalled, who clearly hoped "to have a lot of fun at our expense." Instead Howard gave them something to ponder.[16]

No transcript of Howard's talk has survived. Later Howard described the South Orange meeting as precisely the right "psychological moment" for launching the anti-mosquito crusade.[17] Since Howard's tentative experiments with oil in 1892, knowledge about the biology of mosquitoes, mosquito-borne diseases, and practical strategies of control had dramatically increased. Citizens in South Orange and nearby Summit, New Jersey, had begun to organize anti-mosquito campaigns "even before the agency of mosquitoes in the spread of disease became an established fact." Howard's book was the first work by an American entomologist that sought to relate "the whole mosquito story" to devising a systematic strategy for developing effective mosquito control programs.[18]

By all accounts, Howard did not disappoint his audience. The depth of Howard's knowledge of New Jersey's mosquitoes surprised Miller and his fellow progressives. No other state's mosquitoes are discussed in such detail.

Howard made more than twenty references to New Jersey in his *Mosquitoes*. He described in detail the pioneering research of John B. Smith, Howard's friend and former colleague at the USDA who was New Jersey's state entomologist. Howard encouraged the audience to launch an anti-mosquito campaign. Oiling would provide some relief. Time, money, and public support were needed if mosquitoes were to be vanquished from the Oranges.

The next day the New Jersey and New York papers declared that Howard received a thunderous response. His speech was punctuated with outbursts of applause. "The difficulty," a newspaper reporter waggishly observed, "was to know which was applause and which was a mosquito being killed, for the hall was simply filled with mosquitoes. It was just slap, slap, slap."[19]

At the conclusion of Howard's talk, the South Orange residents organized themselves as the Village Improvement Association. By year's end they raised more than a thousand dollars for the war against mosquitoes. In the weeks after Howard's talk, Miller and his colleagues employed "every scheme of publicity" to advertise their efforts. Neighboring villages and towns launched their own anti-mosquito "town protective and improvement associations." By midsummer, a half dozen of these associations formed a Joint Committee on Mosquitoes. Later Howard maintained that the American anti-mosquito movement began in New Jersey.

The Reed Commission sparked a worldwide interest in mosquito control. In other lands, eliminating mosquito-borne diseases became the primary motivation for the anti-mosquito work. This was the legacy of Ross and the Reed Commission. In New Jersey, Howard argued, a desire for comfort and increased property values triggered what was to become the American mosquito crusade.[20]

While the members of the Village Improvement Association sought to banish mosquitoes from their verandas, eliminating mosquito-borne diseases was very much on the mind of Alvah Doty, New York City's health officer. Doty bore the responsibility of protecting the city from the spread of infectious disease for more than thirty years. In 1878 he had become an inspector for the Health Officer of the Port of New York. He soon became the Board of Health's chief diagnostician.

During the 1890s typhus and cholera epidemics ravaged the city's poor. Newspapers reports blamed recent immigrants for the outbreaks. Pulitzer's *New York Herald* hammered home the point with a cartoon "showing Europe as a giant skull sending forth immigrant ships labeled "Hunger Typhus," "Diphtheria," "Yellow Fever." Respected publications such as the *Boston Medical and Surgical Journal* joined the anti-immigrant chorus declaring, "We open our doors to squalor and filth and misery—which means typhus fever."[21]

In 1895 Alvah Doty was appointed the chief of the Bureau of Contagious Disease for the Port Authority. Doty set about reforming the city's quarantine procedures. He shared the progressive belief that public health authorities had a responsibility to prevent outbreaks of disease. During his seventeen years leading the city's fight against yellow fever, malaria, cholera, plague, and typhus, Doty earned a reputation for his struggle to improve the conditions of the city's poor. In 1907 Robert H. Fuller, the governor's secretary, declared, "Previous to Dr. Doty's original appointment the quarantine regulations had been managed in such a way that New York was often threatened with contagious diseases brought through the port. Dr. Doty's management of infectious and quarantinable diseases has prevented the recurrence of such dangers."[22]

At the turn of the century malaria was widespread in New York City. Local physicians reported an alarming increase in the malaria rate in the boroughs of Manhattan, Brooklyn, and Staten Island. There were several reasons for this. Immigrants from areas in southern and Eastern Europe where malaria was endemic introduced new sources of infection. Poorly planned construction created numerous opportunities for mosquito breeding. On July 31, 1901, the New York Health Board issued an emergency bulletin. "Malarial fever is quite prevalent," the city's health officer declared, " . . . in view of the extensive [subway] excavations and consequent formation of rain pools in various parts of these boroughs."[23]

Three days later Doty announced the Health Department's "crusade against the plague of mosquitoes" on Staten Island. Doty employed forty men to survey a two-square-mile area on Staten Island for pools of standing water. He also appealed to people living in the Concord Basin for help in collecting mosquitoes. Health inspectors supplied "large test tubes with cotton stoppers" to the local population to hold the captured mosquitoes.[24]

The public gave scant attention to Doty's efforts to collect mosquitoes until he offered a ten-cent bounty for each captured mosquito. Within twenty-four hours, he had one hundred specimens.[25] "We have caught the anopheles," Doty triumphantly declared, "the malaria-spreading mosquito red-handed."[26] The Standard Oil Company supplied Doty with one hundred barrels of "Lima oil," called this because the oil was pumped from an oil field in Lima, Ohio, to use in his campaign against anopheline mosquitoes.[27] By August 18, 1901, the anti-mosquito crusade showed positive results. "I have learned," Doty told a reporter from the *New York Times*, "that what I did at Concord has had the best results. Although the district is surrounded by the breeding places of mosquitoes which I could not deal with without going outside my programme, the people in the basin which I experimented with have noticed a marked abatement of the pest. The cases of malaria there have all done well. . . . I have received many communications from all parts of the United States on the subject of mosquitoes and malaria. . . . It is a fascinating study."[28]

Even before Doty's Concord campaign, a handful of progressive public officials and physicians had launched their own tentative anti-anopheline work. The previous year the city council in Winchester, Virginia, passed an anti-mosquito ordinance and issued a public proclamation outlining ways to "Prevent the Mosquito." R. J. Barton, Winchester's mayor, later told Howard that initially the council's efforts to eliminate mosquitoes were "greeted with merriment." Barton reported that he had become the subject of "very sharp ridicule" for his support of the anti-mosquito measures. By October Winchester's campaign had begun to show positive results. "I thank you very much in my own name," Barton wrote to Howard, "and in that of all our people, for the good that your wise and practical investigations have done."[29]

Doty's success on Staten Island encouraged other physicians. In 1901 Dr. A. N. Berkeley organized anti-anopheline measures in several small towns near New York City. A year later Dr. J. M. Barnett launched what was probably the first southern anti-malaria effort based on mosquito control in the small southwest Georgia mill town of Pretoria. The town's clay soil and limestone bedrock made for poor drainage, and malaria was endemic. In 1902 Barnett convinced local businessmen and city leaders to support an anti-malaria initiative. "In a few weeks," Barnett recalled, "every house was screened, every bed netted and all standing water within a half mile of the camp was drained. Larger bodies of water were oiled with crude oil every seven days."[30] Pretoria's malaria rate remained low until the town discontinued the anti-mosquito effort in 1904.

There were opponents to Doty's anti-mosquito measures. The New York papers reported that for the first time a handful of individuals protested against the oil's harmful effects. On Staten Island, farmers and anglers complained that the oil injured both domestic and wild animals. Doty fell under heavy criticism when a self-described "goose woman" protested that some of her geese were dying despite her best efforts in sleepless nights spent scrubbing "from the bodies of her geese the collection of Dr. Doty's petroleum which they industriously accumulate during the day."[31]

Commercial fishermen declared that Doty's anti-mosquito crusade was a mixed blessing. They acknowledged that Doty's oiling had reduced the mosquito problem but charged that the oil had also decimated the fish population. Paul DuPont, a Staten Island native, reported that he had placed a fish cage in one of the ponds that Doty's inspectors had treated. The next day DuPont discovered that all of his fish had suffocated.[32]

Staten Island's poor and immigrant population considered the Lima oil treatments a blessing. Doty reported that the malaria rate plummeted in the weeks following the anti-mosquito measures. For the poor, Doty's campaign had improved their lives. "We are too poor to buy mosquito netting for the windows," a Concord resident told a *New York Times* reporter. "There can be no doubt that the relief came from what has been done here. . . . Last night for the first time this summer we had a lamp in the sitting room. We never dared to do this before."[33]

Across the Hudson River in New Jersey the Village Improvement Association's anti-mosquito campaign also drew criticism because of its use of oil. Miller and his associates soon learned that oil was not a cure-all. Despite repeated applications, mosquitoes flourished in the Oranges. In June, an article in the distant *Los Angeles Times* mocked the New Jersey mosquito crusaders' efforts. Worse, the hardy "Jersey Skeeters" had "succeeded in getting through the oil" and "seem to have fattened and grown more bold and voracious and multiplied alarmingly on it." The "bewildered entomologists," the article concluded, "were at their wits' end to account for this failure." In New Jersey, the article concluded, "mosquitoes *defy* science."[34]

The "mosquito troubled waters of South Orange" grew even more murky when the *New York Times* published an inflammatory front-page story on July 21, 1901, titled "Mosquitoes as Firebugs: New Jersey Report Says They Fill Up on Oil and Menace Property." The reporter quoted Dr. W. W. Heberton, health officer in South Orange, as saying that he blamed the setback on "citizens who do not act in cordial co-operation in the war of extermination." Some objected to the smell and refused to apply oil. Others applied too little. Worse, some applied too much. In South Orange, homeowners were warned to be on the lookout for oil-saturated "blazing mosquitoes" that "have set fire to curtains and draperies before the insects were consumed."[35]

The nascent mosquito crusade in northern New Jersey might have ended at this point if it had not been for the leadership of John B. Smith, Howard's former assistant. In 1889 Smith left the USDA to become the entomologist at New Jersey's Agriculture Experiment Station at Rutgers. Between 1901 and his death in 1912, Smith laid the foundation for the development of mosquito control in the twentieth century.

Born in 1858, Smith grew up in a family that was passionately interested in insects. Smith's father, who changed the family name from Schmidt to Smith upon arriving in America, was an amateur entomologist and cabinetmaker.[36] In Germany, he had invented the "Schmidt Box," a cigar-box like container for insects. The Schmidt Box was a standard entomological collection tool well into the twentieth century. Believing that there were few opportunities for entomologists, Smith's father encouraged his son to study law. In 1879 Smith passed the New York bar exam and launched his legal career. Friends reported that Smith once observed, "a fly on the wall was more interesting than the case in hand."[37] Indeed, his passion lay in the study of insects. In 1882 he found an outlet for his entomological interests when he joined the Brooklyn Entomological Society. Smith and George D. Hulst, a prominent amateur entomologist and pastor of a local church, became close friends. Smith shared Hulst's fascination with lepidoptera and became a regular contributor to the society's journal and eventually the journal's editor for two years.

Smith's legal career ended in 1883 when he joined the USDA's division of entomology as a field agent. His first assignment was to prepare a study of the insect pests that were damaging cranberries. Two years later, Howard recommended Smith for a position at the National Museum as the assistant curator of the division of insects.

While Smith was cataloguing insects at the National Museum, his friend George Hulst had left the pulpit and accepted a position at Rutgers. In 1888 Hulst became Rutgers's first entomology professor. After one year Hulst decided that he was better suited to saving souls than teaching undergraduates. He recommended that his thirty-one-year-old friend, John B. Smith, be named his successor. On April 1, 1889, Smith reported to New Brunswick. He served as the New Jersey Agricultural Experiment Station's entomologist for the next twenty-three years.

Self-taught and possessing no formal scientific training in entomology, the newly appointed professor committed himself to seeking practical solutions to insect problems. Throughout his career Smith cultivated his relationships with the farmers, nurserymen, beekeepers, and horticulturalists throughout New Jersey. One of his first acts was to publish a special bulletin calling for suggestions of topics that needed study. Within a few weeks Smith had assembled an impressive roster of entomological questions. In his early work Smith continued Hulst's effort to build a collection of notable New Jersey insects. By the end of his first year at Rutgers, Smith had prepared studies of the horn fly, plum curculio, and other economically significant pests. By the mid-1890s Smith had conducted pesticide experiments involving London purple and Paris green (two arsenic-based poisons), kerosene emulsions, and carbon bisulfide, as well as other chemicals.

Smith gained national prominence in 1896 for his efforts to control the San Jose scale. In 1894 an unprecedented statewide infestation of New Jersey's apple and pear orchards by the San Jose scale produced a severe economic loss. Entomologists first identified this fruit tree pest in San Jose, California, in 1870 in a shipment of plants from the Far East. Smith secured a thousand-dollar special appropriation from the State Legislature, which financed a research trip to California to research the San Jose scale's natural enemies.

When he returned from California he organized a program to protect New Jersey's apple and pear orchards, which relied on chemical (spraying trees with edible fish oil) and natural or biological control by establishing colonies of ladybugs.

The earliest evidence of Smith's growing interest in mosquitoes appeared in September 1896 in the midst of his campaign against the San Jose scale. Smith prepared a monthly article on economic entomology for American Academy of Natural Science's *Entomological News*. "Mosquitoes," Smith wrote in September issue, "were also the subject of much newspaper comment and in some places near New York they were certainly numerous." Smith devoted his remarks to explaining the increased numbers of winged pests ("heavy rains," "choked ditches," and "hot weather"), the growing mosquito problem in California due to increased irrigation, and the ability of mosquito eggs to survive in dry periods. "New Jersey has an enviable reputation for the quality of its mosquitoes," Smith declared tongue-in-cheek, "but I believe that the foreign product; on the plains of Manitoba, is superior in size, and at least equal in numbers and blood-thirsty disposition."[38]

Like Howard at the USDA, Smith received numerous inquires about mosquitoes in the late 1890s. Berry pickers refused to work. Innkeepers on the Jersey Coast blamed mosquitoes for cancelled reservations and empty hotel rooms. Officials in Atlantic City acknowledged that when the mosquitoes were on the wing, only the most intrepid dared to venture on the newly constructed boardwalk. Homeowners in South Orange, Elizabeth, Newark, and Jersey City regularly wrote to Smith chronicling the depredations of the bloodthirsty pests. "New Jersey's reputation for mosquitoes is well established," Smith wryly observed, "and more people come into our State annually to be bitten by our shore species than go to any other State in the Union for like purpose."[39] Smith made it his business to do something about it.

Smith first outlined his plans for a comprehensive anti-mosquito program in the "Report of the Entomological Department of the New Jersey Agricultural College Experiment Station for the Year 1901." "I can see no reason why, inside a decade," he declared in the report's conclusion, "New Jersey mosquitoes should not be reduced to such a point as to be practically unnoticed. Extermination, be it noted, is *not* claimed. The plan contemplates permanent relief, based upon intelligent activity, along well developed lines."[40]

Smith's active research on mosquitoes began in the late 1890s. In 1900 Smith persuaded Edward Voorhees, the director of the state Agricultural Experiment Station, to provide him with a small sum of money to research New Jersey's mosquito problem.[41] The next year in his annual report on the entomological department, Smith announced his intention to develop a "rational method" of mosquito control focusing on the life history of mosquitoes and "the natural checks and methods by which they may be made most effective."[42]

Smith distanced himself from Howard and those who believed that oiling offered an easy solution to the mosquito problem. Anticipating the environmental concerns of later generations, Smith opposed the wholesale use of "destructive" chemical means of control. "The treatment of running streams or bodies of water in which fish, frogs, or other aquatic animals live, or can be made to live, is unnecessary, and even in some cases absurd," Smith declared, "particularly if the agent employed is destructive to fish." It was more effective to stock a pond with the "proper fish" than "spray it periodically with oils." For Smith, "the first essential in my plan of control . . . is to ascertain all the natural checks."[43]

Smith's call for a "rational method" of mosquito control came at a critical moment for Progressive forces in New Jersey. In 1899 a scandal erupted over the Hudson County almshouse that pitted the local political machine against a reforming governor and his supporters. The controversy surfaced when a muckraking reporter for the *New York Herald* revealed the horrific conditions at the almshouse. Hudson County's almshouse, lunatic asylum, and penitentiary were located in a desolate part of the Hackensack salt marsh that was notorious "for the most voracious breed of mosquitoes in the country." The subsequent investigation of conditions at the almshouse revealed it to be "an abode of filth" and degradation where county officials profited "from the sale of parentless youngsters" to private adoption companies.[44]

Foster Voorhees, New Jersey's governor, asked Emily Williamson to lead the newly created Board of Children's Guardians. Williamson, an advocate of women's suffrage and other Progressive reforms, set to work correcting the abuses at the almshouse. "Reclamation," William Sackett wrote a decade later, "was the password to the other legislation of the Voorhees regime. While Mrs. Williamson was exerting herself to save its citizenship . . . the Governor was interesting himself in the other direction of developing its natural resources." Voorhees directed the state's engineer, C. C. Vermeule, to prepare a plan "for the reclamation of the miles of waste meadows lying between the west slope of Bergen Hill in Jersey City and the Passaic River."[45]

Eliminating the mosquito blight became a central component in the plan to improve the value of the Hackensack marshes. William T. Hoffman, a reforming Hudson county judge and newly elected member of legislature, called for a war against mosquitoes. Hoffman, whose colleagues called him the "mosquito statesman," declared that "thousands of acres that ought to be priceless were made uninhabitable at times when they should be most sought."

Hoffman argued that eliminating mosquitoes "would add inestimable millions to property values."[46]

Smith's reports on the state's mosquito problem could not have come at a better time. Progressive politicians supported the anti-mosquito campaign for both economic and health reasons. In November 1901 voters in Jersey City elected Mark Fagan, a thirty-two-year-old reformer and former assistant undertaker, mayor. In the weeks before Election Day Fagan promised to improve schools and quality of life for the city's poor and immigrant population. Ridding Jersey City of mosquitoes would allow people to enjoy parks and evening concerts. "I wanted," Fagan explained, "to make Jersey City a pleasant place to live in; I'd like to make it pretty."[47]

The Progressives' successes convinced Smith that there was support for expanding his mosquito study. In 1902, after working alone for nearly two years, Smith submitted a request for ten thousand dollars to the New Jersey legislature. Smith's objective was to "present a consistent plan of action, which will show what can be done by the individual, the local community and the State—a plan which will attain the result slowly or rapidly, in proportion to the energy or listlessness, with which it is carried out and which will prevent discouragements and failures due to attempts to accomplish the impossible."[48]

On April 3, 1902, the New Jersey Legislature approved Smith's request for ten thousand dollars "to investigate and report upon the mosquitoes occurring within the state, their habits, life history, breeding places, relation to malarial and other diseases, the injury caused by them to agricultural, sanitary and other interests of the state, their natural enemies, and the best methods of lessening, controlling or other wise diminishing their numbers."[49] The appropriations committee, however, neglected to include funding for the survey in its budget. Smith appealed directly to the newly elected governor, Franklin Murphy, an advocate of the mosquito crusade, who supplied Smith with one thousand dollars from his emergency fund so that he could immediately begin work.

When Smith opened his investigation, entomologists knew little about mosquitoes. There were no systematic catalogues or keys of mosquitoes to guide him. Earlier researchers had focused almost exclusively on *Anopheles* mosquitoes, the malaria vector. For Smith, Howard's "little book [was] the only tangible starting point."[50] Smith set himself three tasks: first, to identify the mosquito species that were most troublesome in New Jersey; second, to provide a detailed examination of the life history of each of these species; and, third, to assess the practical methods of eliminating those species of mosquitoes that posed a challenge to New Jersey's comfort and health.

WAR ON THE BLOOD SUCKERS.
New Jersey Inaugurates a Crusade Against the Merger and the Mosquito.

Figure 1.1. J. N. "Ding" Darling editorial cartoon, "War on the Blood Suckers," March 8, 1902. Reproduced courtesy of the "Ding" Darling Wildlife Society.

Smith's scientific work rested on his scrupulous attention to detail, his ability to reach general conclusions, and his knack for identifying practical solutions. Friends respected and his enemies feared his tenacity in pursuing what he judged the best course of action. These qualities contributed to what was Smith's first important discovery.

Shortly after Smith began his research on mosquitoes, prodigious numbers of *Aedes sollicitans*, locally known as the white-banded salt marsh mosquito, descended on New Brunswick. "My garden," Smith later wrote, "was full of them, and there was no remaining among the shrubs and bushes when the sun began to sink. Nevertheless, I could get no larvae of the species. I prepared pails in all sorts of ways and even added sea-salt to some of the water to induce oviposition." None of Smith's efforts to produce larvae worked. An expanded search produced no evidence of *Ae. sollicitans* larvae. "Full of conviction that mosquitoes must breed near where they were found, I hunted every pool near by and along the shores of the Raritan river . . . but never found *sollicitans*." By the fall of 1902 Smith recognized his error. He announced that he had conclusive evidence that the white-banded salt marsh mosquito had a flight range of forty or more miles. The common belief that mosquitoes do not fly far was dead wrong.[51]

Two-thirds of New Jersey fell within the flight range of salt marsh mosquitoes. If the Garden State's progressives united, Smith was confident that salt marsh mosquitoes could be controlled. "It is," Smith declared, "quite within our power to deprive [*Ae. sollicitans*] of their larger breeding areas and to reduce them to a point where they will cease to be obnoxious."[52]

Smith organized a statewide anti-mosquito crusade between 1901 and 1903. He crisscrossed the state meeting with local health boards, municipal officials, physicians, sanitarians, engineers, lawyers, and teachers. Smith later wrote, "I have lectured before associations of many sorts, and I believe that as the result there is at present much better knowledge of mosquitoes than ever before."[53]

There were, Smith explained in subsequent publications, two approaches to mosquito control: destructive and permanent. Destructive mosquito control sought to destroy mosquitoes either by encouraging their natural enemies or by introducing an agent such as oil that killed mosquitoes. Permanent or preventive mosquito control eliminated larval habitats. Smith contended that the solution to New Jersey's mosquito problem lay in permanent control.

In 1903 Smith finally secured a recommendation from the appropriations committee that nine thousand dollars be allocated for the mosquito study. Smith was becoming an adept politician. "I wrote you," he confided to his friend Turner Brakeley, "I was going to Trenton to see the Committee on Appropriations and I did. I carried with me some of the last filled bottles of your bog born babies as an object lesson and it impressed them largely. They had never seen wrigglers before and the idea that these miserable things were already laying for them was a shock."[54] Smith's "bog born babies" proved a telling argument in the state capital. "The Legislature of the State of New Jersey adjourned last night," Smith wrote to Leland Howard. "Peace to its ashes. It was a real good Legislature. It treated me very decently and only cut one thousand dollars from my appropriation." The bill authorized Smith to spend five thousand dollars before the current fiscal year's end on October 31. The remaining four thousand dollars were set aside for the following twelve months. "I need hardly say to you," Smith told Howard, "that I expect to spend every cent of the appropriation."[55]

The appropriation meant that Smith had the resources to develop a comprehensive anti-mosquito program. Smith confided to Howard the outlines of this program. "I feel that I owe you," Smith wrote, "some consideration in this matter, and the least I can do is to tell you what I hope to do and what I do not expect to try to do." Smith first explained that he would limit his "work entirely to the faunal district to which the State of New Jersey belongs." Then he declared, "I have no intention of carrying on any further experiments concerning the relationship of any species of *Anopheles* to the distribution of malaria." Smith's objective goal was to identify the species of mosquitoes that were troublesome in New Jersey and to devise a practical plan of permanent control. "It is not part of my plan of work," he concluded, "to do systematic work with the species of mosquitoes."[56]

Smith had recruited a cadre of assistants before receiving the state funds. In 1902, well before he could "claim a practical outcome for the study," Smith and Edgar Dickerson, his assistant in the entomology department at Rutgers, took the first steps in working out the life history of *Ae. sollicitans*.[57] Governor Voorhees's emergency appropriation allowed him recruit additional help. "I think," Smith wrote in March to Hermann Brehme, a self-employed dealer in all kinds of insects and entomological supplies "we are about ready to start mosquito work and I want a little conversation with you on the subject."[58] Brehme's assignment was the salt marsh problem. Brehme's diligence, precision, and unwavering devotion to the mosquito crusade made him Smith's most valuable worker.

Smith chose John Grossbeck, Henry Vierreck, and William Seal as his field agents. Their charge was to identify oviposition sites, survey larval habitats, collect specimens, and execute Smith's various control experiments. Grossbeck covered the state's interior, gathering information on freshwater mosquitoes. Smith sent Vierreck to Cape May County at the southern tip of the state. Smith made Seal, who had retired from the U.S. Fish Commission, responsible for gathering information on the natural enemies of mosquitoes. Under Smith's direction Seal traveled to North Carolina and brought back *Gambusia affinis*, top water minnows, which Smith hoped to use as a means of natural control.

Disaster struck barely three weeks after Smith received the legislature's appropriation. "We indulged in the luxury of a fire," Smith wrote, "about a week ago and while the entire collection was saved a great many of my papers were destroyed or rendered useless."[59] When Smith learned of the fire, he rushed to the entomology building to find his students trying to save his insect collection. Before Smith's "room was invaded by fire [the students] cleared out almost every last box before the smoke and flames on the stairway stopped work." Smith was a stoic. "My loss has been on books, manuscripts, and records.

Figure 1.2. John Smith and Leland Howard circa 1901. Photo courtesy of the Center for Vector Biology, Rutgers, The State University of New Jersey.

Books of course can be purchased again. Tools etc. are covered by insurance. The notes, or rather their loss, will inconvenience me very considerably but they can be done over again. On the whole I got off very well."[60]

Word of the fire spread through the entomological community. Smith received numerous offers of help. Insult was added to injury, however, in early May when Smith received a letter from Harrison Dyar, curator for insects at the National Museum. Dyar, notorious among his contemporaries for his caustic temper, wrote inquiring about the collection of lepidoptera that George Hulst had entrusted to Smith when Hulst returned to Brooklyn. Dyar chided, "You ought to have put these types in the Natl. collection long ago as I have repeatedly urged before this warning was directed against you. In the next fire you may not fare so well."[61]

Three months earlier Dyar used his farewell address as president of the Entomological Society of Washington to criticize Smith's scientific work as "vague" and "discursive." Smith's "synoptic tables . . . [are] somewhat overdone. I would not say that the characters used are sometimes imaginary, yet they verge upon this definition." Dyar reserved his severest censure for George Hulst, Smith's friend and benefactor. "Nothing that Hulst has done can be absolutely relied upon," Dyar opined, "for fear that the thing, apparently most evident, may be found to be vitiated by some blunder he knew much better than to commit."[62]

In early April Smith's and Dyar's relations reached a breaking point when Smith learned Dyar had published a description of mosquito larvae that Smith had discovered and sent to the National Museum. When Smith protested, Dyar laconically responded, "I shall probably never get used to your extremely personal point of view."[63] A few weeks later Smith discovered that Dyar had secretly contacted John Grossbeck. Dyar wanted Grossbeck to send him samples of the mosquito larvae Grossbeck had collected for Smith. Smith was furious when Grossbeck revealed Dyar's request. Dyar had gone too far. Smith vowed to have no further relations with the National Museum so long as Dyar remained in Washington. "Not one specimen of either the Hulst or my own collection," Smith wrote to Dyar, "will go to the National Museum."[64]

Smith had more to worry about than the fire and his quarrel with Dyar. On May 11 Spencer Miller sent Smith a telegram asking for help. "South Orange full of mosquitoes, wind from East, board of health meets to-morrow night. Can you or your representative visit the village to-morrow and report to the Board of Health."[65] Smith had become directly involved in the association's anti-mosquito work in early 1903. Smith had written Miller and offered to assist the Improvement Association. "I am anxious," Miller responded, "to avail myself of your kindness."[66]

Smith sent Hermann Brehme to South Orange to survey the salt marshes. While Brehme prepared his report, the association applied oil to the ponds, cisterns, and marshes within the village's boundaries. The members had divided the town into four mosquito districts. Each district was a semi-autonomous entity responsible for raising money, securing oil, hiring workers, and treating the waters within its jurisdiction. The system worked in the well-to-do districts. The poorer parts of the community were unable to raise money to purchase oil, leaving the mosquito problem unabated.

"The anti-mosquito machine in South Orange, a reporter for the *New York Times* declared, "has sprung a leak and it will take several days to repair the damage, for even the fighters themselves admit that their forces are inadequate to meet the phalanx of mosquitos that has come with the recent rains."

The *Times* declared that critics of the mosquito crusade were saying "unkind things about the oil lately, claiming it attracts the mosquitoes instead of repelling them."[67]

Miller and his allies in the Village Improvement Association feared that the clouds of mosquitoes and the newspaper stories would discredit the anti-mosquito movement. Later in the day, he vented his frustration in a handwritten note to Smith. "I am met everywhere and on every side with this complaint. . . . It is very discouraging, in fact so much so that it will be difficult to raise any more money to carry on the work."[68]

Miller and his wealthy associates' exasperation grew out of a fundamental misunderstanding of the mosquito problem. The South Orange Improvement Association based its campaign on eliminating mosquito breeding within the township. They failed to grasp that there were, in fact, two distinct problems. The first consisted of mosquitoes with short flight range. The second far greater problem came from the salt marshes, which bred *Ae. sollicitans*. These mosquitoes had a flight range extending over most of the state.[69]

South Orange's mosquito predicament brought into focus the challenge facing the mosquito movement. Part of the problem was indeed purely a local matter. The state's salt marshes, though, were a different matter. They covered roughly 296,000 acres of tidal plane.[70] Smith knew that the salt marsh problem was beyond any single community's ability to respond. The current problem in South Orange illustrated this. "I know exactly where the South Orange mosquitoes come from," Smith wrote to Dr. W. W. Heberton, secretary of the local health board, "There are untold millions of them on the Newark meadows, and in fact all along the shore line as far as Perth Amboy. . . . Your work has been entirely satisfactory. I have had two men working in your district for over a week and there is not trouble with what you have done. This occurrence is one of those that must be expected so long as the Newark marshes remain untreated."[71]

Reformers rallied support for mosquito control. "In wetlands," historian Ann Vileisis has noted, "progressive local governments found one of their earliest successes. . . . Armed with scientific authority, recently organized city and state health boards spearheaded drainage and mosquito eradication programs."[72] Smith's work supplied New Jersey progressives with the scientific authority they needed. Moreover, he had already set to work on solving both elements of the mosquito problem. He was confident that something could be done immediately to help local groups. A first step was organizing the various anti-mosquito associations. In 1903 Smith organized the Conference Committee on Mosquito Extermination.[73] The committee consisted of representatives from the various progressive citizen associations and local governments that had joined the anti-mosquito crusade. The conference meetings served as a forum in which the members could exchange ideas and discuss common problems.

The relationship between the anti-mosquito movement and local health boards was frequently discussed. In general, local health boards had responded to Smith's call for help. Newark's and Elizabeth's health boards were notable for their eagerness to address the mosquito problem. Elizabeth was the first large city in New Jersey to take action against mosquitoes. A local physician, William Robinson, opened the campaign when he delivered a talk on the mosquito problem in 1902. Robinson's speech led to the formation of a committee of sixteen business and community leaders. Each agreed to contribute twenty-five dollars to the crusade. The citizen group had raised five hundred dollars. In April 1903 Elizabeth's city council allocated a thousand dollars for continuing the anti-mosquito work. The council asked Louis Richard, the city's health officer, to supervise the work. Similarly, in nearby Newark the local health board took the lead in organizing the anti-mosquito movement.[74]

In 1903 they approved an "experiment with machine ditching" on the salt marshes that lay within Newark's city limits.[75] The city's ditching effort proved futile because of its limited extent. A decade later a resident recalled that there were times "when mosquitoes came into Newark and Jersey City and penetrated the stores in such numbers that they had to be swept out of the show windows after they died."[76]

Smith applauded the initiatives of Newark's and Elizabeth's reform-minded health boards. Their "experiments" were useful first steps in local control. The larger problem lay in the salt marsh (now the Newark Liberty International Airport) that separated the cities. Ridding the cities of mosquitoes meant expanding the local health boards' ability to implement anti-mosquito measures.

Not all health boards were willing to join the anti-mosquito campaign. There was general agreement that health boards should act against *Anopheles* mosquitoes because of their role in the transmission of malaria. Some health officers and boards of health considered questionable whether other mosquito genera, no matter how annoying or indirectly dangerous they might be, posed a threat to public health. Accordingly, these boards were hesitant in taking action against mosquitoes.[77]

This narrow interpretation of what constituted a threat to public health undercut the work of even progressive health boards. If this notion spread, the advance of the anti-mosquito movement would be stopped. In fact, this occurred before the end of 1903. A property owner in Newark won an injunction against the city's anti-mosquito initiative, claiming that the health board had no right to trespass on his property because the mosquitoes breeding on his land posed no health threat.[78]

Smith thought that this problem could be solved by expanding the 1887 New Jersey Health Act to include an explicit mandate ordering health boards to act against mosquitoes. The challenge was twofold: first, agreeing on the proper language, and second, mobilizing political support for the proposal. Edward Duffield, one of the original members of the South Orange Village Improvement Association, supplied both.

Duffield's commitment to reforming the General Health Act grew out of a promise he made to Spencer Miller. In 1902 the Essex County Republican Party asked Miller to run for the New Jersey Assembly. When Miller learned that Duffield, a local attorney and a member of the Village Improvement Association, wanted to run, Miller dropped out of the race and supported Duffield's candidacy. Duffield visited Miller after the election to thank him for his support. "I know your passion," the newly elected assembly member declared. "I know what you want to do and if there is any way I can help you in the legislature on your great undertaking, I want to help you." In 1903 Miller asked for Duffield's help. "I want a law passed that a man who breeds mosquitoes on his premises is creating a nuisance and is subject to the board of health restrictions for those nuisances."[79] Duffield studied the General Health Act and proposed the addition of the six words "waters in which mosquito larvae breed" to the sections of the law defining public nuisances.

Winning approval of the Duffield Amendment was not a foregone conclusion. The amendment had powerful opponents. Chief among them was Henry Mitchell, the conservative secretary of the state health board. At first Smith tried to persuade Mitchell to support the amendment. "I was in Trenton yesterday and in Trenton one learns all sorts of peculiar things," Smith wrote to Mitchell at the end of February 1904. "One of the most peculiar was that you were antagonizing the amendment to the general health laws."[80] Smith pointed to the example of how a single property owner had blocked Newark's health board.

Mitchell was unconvinced. "It is wholly unnecessary to amend the law as you propose," he replied to Smith. "Furthermore it is inadvisable, in my opinion, to needlessly alter the act."[81] He maintained that local health boards already possessed whatever power they needed for anti-mosquito work under the 1887 General Health Act. Smith fired back, "Lawyers do not agree with your view of the case." The General Health Act authorized local health boards to abate only pools containing *Anopheles* larvae. "Mosquitoes aside from the species of *Anopheles* are not necessarily injurious to public health," Smith wrote, "and I personally doubt very much indeed whether courts would sustain any local board in action under the present general powers if there was persistent opposition." It was imperative that the Duffield Amendment pass. The future of the mosquito crusade was uncertain.[82]

Smith's proficiency as a politician improved with each legislative session. The New Jersey exhibit at the 1904 Louisiana Purchase Exposition in St. Louis demonstrates this. While the Duffield Amendment moved through the legislature, Smith quietly made certain that a display on mosquito extermination was included in New Jersey's pavilion. New Jersey long bore the dubious reputation of being the "Skeeter State." In 1903 the Edison Company had distributed nationwide a comic film called *Smashing a Jersey Mosquito*. In the short, a woman and her husband are sitting in their comfortable home when a gigantic mosquito attacks them. The man seizes a broom and tries to kill the pest. Mayhem ensues. Mirrors, pictures, chairs, tables, a cuckoo clock, and various sundry things are destroyed in the couple's futile effort to kill the pest. Finally, the woman wounds the mosquito. Her husband jumps on the bloodthirsty beast. The film ends with a terrible explosion producing "great clouds of malarial gas" which envelop the once-happy New Jersey home.[83]

Confounding these stereotypes was one of Smith's objectives in his mosquito extermination exhibit. The completed display demonstrated the work that New Jersey had done to control the mosquito pest. Cabinet displays showed the various stages in the life history of New Jersey's principal species of mosquitoes. Smith placed a model city at the center of the pavilion. There was a marsh at the end of the city's square. One side of the marsh represented ideal mosquito habitats. The other illustrated how progressive health boards and improvement associations were using both drainage and the mosquitoes' natural enemies to control the pest. The message was clear: Nature could be corrected. Smith's engineered landscape held out the promise of making New Jersey a "Skeeterless State."

Whether the mosquito extermination exhibit succeeded in correcting Edison's film version of New Jersey is uncertain. The exhibit did pay a sizeable dividend in Trenton. Smith included portraits of the local and state officials who had supported the anti-mosquito movement. He sent letters requesting photographs to prominent individuals throughout the state. "As Governor Murphy assisted all efforts in that direction [mosquito extermination] from the beginning," Smith wrote, "I would like to show his photograph among those entitled to credit."[84] To friends like Spencer Miller, Smith joked, "I wish to declare emphatically, however, that I do not want your picture as a help to exterminate mosquitoes."[85]

The battle over the Duffield Amendment came to a head in late February and early March. Smith wrote to both the Assembly's speaker and the president of the New Jersey Senate. "I am writing by the same mail," Smith explained in his letter to Senate president Edmund Wakelee from Bergen, "to Speaker [John] Avis . . . suggesting to him that if it is not entirely out of line, I would like to present to the legislature a talk of about an hour, the results of work done by me in the mosquito investigations."[86]

The slide lecture was a success. "The ways of the legislators," Smith confided to Eugene Winship, a leader in both the Rumson Neck Mosquito Extermination Association and Monmouth Beach Improvement Association, "are past finding out. I am a little afraid that whatever fight is made against the bill will be made before the Governor."[87] Smith was right. His frequent visits to Trenton succeeded in convincing the Legislature that "mosquito work is [was] not a fool proposition, but a very vital one of the interests of New Jersey."[88] The Duffield Amendment passed both houses on March 11 and was sent to Governor Franklin Murphy.

Three days later Smith wrote to Murphy urging him to sign the bill. The objective, Smith explained is "to make pools of water breeding mosquitoes a nuisance." There were no other changes in the law. He assured Murphy that the Duffield Amendment did not compel local health boards to pass anti-mosquito ordinances. It merely gave health boards the option to do so if they deemed it necessary. Smith saved his strongest argument for the letter's conclusion. Governor Murphy was from Newark. He should know that work had started in Newark in 1903 but that "it was held up last year by a factious opposition of one or two landowners." Smith wagered that a Progressive governor would champion the anti-mosquito cause. He was not disappointed. Murphy signed the bill into law on March 30.[89]

Much had happened in the twelve months since the fire at Smith's Rutgers office. The entomology department had found a new home. "The Experiment Station," Smith explained to Byron Cummings, a Rutgers graduate who was a professor in Salt Lake City, "is now occupying the Fine Arts Building, having driven out all the ancient Gods and some of the Goddesses."[90] His report on the mosquitoes of New Jersey was nearly completed. He planned to publish it at the beginning of the next legislative session. In February, under Smith's direction, Eugene Winship and Hermann Brehme began a drainage experiment in the marshes along the Shrewsbury River at Monmouth Beach and Rumson's Neck.

The initial field reports were positive. September brought confirmation of the effectiveness of Smith's plan. "Mosquitoless Monmouth," Winship wrote, "is the appellation now used: health and comfort have been obtained at a very small cost, considering the benefit derived. The advance in real estate and the increased number of rentals, caused by the absence of the mosquito, will surely be an argument that none can combat."[91] A few thousand feet of ditches along the Shrewsbury River was merely an experiment. The real work remained. In July Smith sent Brehme to Atlantic City and on to Tuckerton to survey the salt marshes. "What did I ever do to you that you sent me over such a confounded road," Brehme wrote. "We did not use horses to pull the stage; *Culex* [*Aedes*] *sollicitans* did that; their pressure on the stage made it run all right. My newspaper was covered with blood from killing them. But we are here and will stand the battle."[92]

Indeed, Smith had reason to celebrate in 1904. Local health boards now had a tool that gave them power to act against mosquitoes. This was an important advance. Unexpected challenges and opportunities lay ahead. In the next seven years, Smith would fight against two-legged adversaries, join forces with Alvah Doty in guiding anti-mosquito work in New York City, inspire the beginning of the mosquito crusade on the Pacific Coast, and wage an unrelenting war against the myriad "Goddesses," the "bog born babies" of the Jersey Shore.

Chapter 2 The Garden State Takes the Lead

I AM IN RECEIPT OF your report on Mosquitoes," a young Berkeley professor named Henry J. Quayle wrote to John Smith at Rutgers in April 1905, "and wish to thank you very much for the favor. I am constantly referring to it in my work on mosquitoes here and find it a great help. Indeed, it is the only guide I have for practical control work on a large scale."[1] Smith's *Report on the Mosquitoes of New Jersey* provided a scientific and practical guidebook for the growing mosquito crusade. "The report," Leland Howard declared, " . . . fills me with delight. It marks a tremendous advance in our knowledge of mosquitoes, and I assure you that it will be of the greatest service to me in many ways."[2]

A cascade of congratulatory letters and requests for assistance arrived at Smith's office in the weeks following the report's publication in February 1905. A real estate developer named Davis in Orlando asked for details on Smith's ditching experiments on the Newark meadows.[3] Delos Lewis Van Dine, the entomologist at the Hawaii Agricultural Experiment Station and organizer of the "citizens' mosquito committee" in Honolulu, inquired if Smith "would . . . be able to recommend a man to come out here."[4] But Quayle's letter was by far the most significant. Three thousand miles west of the Jersey salt marshes, Smith's insights guided Quayle as he fashioned California's first permanent mosquito control program.

Mosquitoes had long plagued the San Francisco Bay area. Early settlers commented on the fierceness and prodigious numbers of blood-sucking pests that appeared each spring. In April 1772 Father Juan Crispi noted in his journal that "vast swarms of mosquitoes" had "sorely afflicted" him as he sought to evangelize the natives near Warm Springs in Alameda County. Forty years later, Spanish soldiers passing through El Cerrito in June complained of being "attacked by swarms of voracious mosquitoes."[5]

Until the beginning of the twentieth century, the residents of Alameda, Marin, San Francisco, and San Mateo counties could do little more than stoically endure the annual spring mosquito infestations. Most people in northern California considered mosquitoes to be "like sunshine and rain, as something beyond our control."[6] In 1903 news of Smith's anti-mosquito experiments stimulated interest in mosquito control in San Mateo County. That spring a group of progressive business leaders formed the San Rafael Improvement Association. Like their counterparts in South Orange, the San Rafael Association was comprised of well-to-do property owners seeking relief from pest mosquitoes. They asked Charles W. Woodworth, who was the "first trained entomologist to assume teaching duties in California" when he joined the University of California in 1891, to advise them in their anti-mosquito work.[7] Woodworth visited San Rafael and discovered that "practically all of the difficulty was due to a single species, *Culex squamiger* [*Aedes squamiger*, a Pacific Coast salt-marsh mosquito], 99% of the mosquitoes being of that species."[8]

In 1901 Woodworth had learned of Smith's work while on sabbatical leave at Harvard where he completed his doctorate. "Professor John Smith, who has done more perhaps in the study of mosquitoes than any other man in America," Woodworth later wrote in a report summarizing the San Rafael work, "had previously made out the very remarkable fact, that one of the commonest and most troublesome mosquitoes in New Jersey was strictly a salt-marsh insect."[9] Woodworth was convinced that Smith's New Jersey work was applicable in San Rafael. His survey of San Rafael's salt marshes confirmed Smith's conclusions. If the members of the San Rafael Improvement Association were to succeed in eliminating the mosquito menace, then they must direct their efforts against the salt marsh pest.

In 1904 Woodworth recommended that the improvement association hire a mosquito control inspector to wage war on the salt marsh mosquitoes. During the previous year Woodworth's students had "repeatedly visited that region" and "made observations as to the distribution of the mosquitoes, collecting specimens of the larva from every part of the town and marshes adjacent, and breeding out the adult mosquitoes, . . . so that the species could be positively determined thus establishing very accurately the range of the breeding grounds." Under Woodworth's supervision the inspector, an unemployed Congregationalist minister, applied oil to the town's salt marshes. One year after his first visit to San Rafael, Woodworth reported that the oil treatments greatly reduced the mosquito problem. After "a half day of hard collecting work," Woodworth declared, "I was able to find less than a score of mosquitoes of any species."[10]

The San Rafael Improvement Association's success inspired the residents of Burlingame, then an exclusive retreat of San Francisco's wealthy elite, to launch their own anti-mosquito initiative. In May 1904 members of the Burlingame Improvement Club sent an "urgent request" to Woodworth appealing for help in fighting mosquitoes, which had "become exceedingly abundant and annoying."[11]

Woodworth dispatched his new assistant Henry Quayle to survey the situation. Quayle, who joined Berkeley's newly formed division of entomology the previous year after graduating from the University of Illinois, spent three weeks tramping through the salt marshes along the west side of San Francisco Bay.[12] He discovered that 95 percent of the problem came from the nearby salt marshes. At the time of his visit, the infestation was so intense that Quayle was able to collect a pint of *Culex currii* [*Aedes dorsalis*] mosquitoes using a net "in less than five minutes." Quayle delivered his report to the club's executive committee in midsummer. He told George Pope, a millionaire lumber tycoon and the club's president, that it was too late in the current season for effective action.[13] Permanent relief from pest mosquitoes, however, was possible if they were "earnest" and prepared to take on "the expense necessary to control this large breeding ground."[14]

Two months passed before Woodworth and Quayle heard back from Pope. Since the June survey, the mosquito problem had intensified. At twilight when the "bog born babies" took flight, wealthy matrons fled indoors, and only the bold ventured out for an evening stroll. "In walking along the roads in the vicinity of the hills, mosquitoes would gather so abundantly on one's clothes that sometimes the color of the suit was obscured beneath the general affect of the light brown produced by the mosquitoes," an observer declared.[15] The situation was intolerable. "As they put it," Quayle later told Smith, "it was up to them either to get rid of the mosquitoes or move out."[16]

In August Pope promised Woodworth and Quayle that the club would finance Quayle's effort to vanquish mosquitoes from Burlingame. In return Quayle promised to begin work in 1905. His first step was to win approval from the university's curriculum committee for what was probably the first college course devoted to pest and nuisance mosquito control. When classes resumed in January 1905, six students had enrolled in Quayle's survey of San Francisco Bay salt marsh mosquito problem. The students soon discovered that this was not going to be a traditional Berkeley seminar. Instead, Quayle took them to Burlingame and set them to work laying out ditches, designing levees, and making plans for refurbishing dikes.[17]

Smith's annual reports on his mosquito work in New Jersey served as Quayle's guide as he drafted his plan. By 1902 Smith had concluded that "ditching is the method by which bad breeding areas are to be redeemed and made harmless."[18] A year later Smith's experiments on the Newark marsh employing the True Ditcher, a mechanical ditching machine, demonstrated that a network of ditches six inches wide and a full two feet deep would "drain the meadow perfectly in twenty-four hours, no matter how heavy the flood."[19] Smith believed that his salt marsh research had proved five things: "First, that salt meadows similar to those in Newark can be readily ditched by machine; second, that the ditches so cut drain the land perfectly for the purpose of preventing the development of mosquitoes; third, that the narrow, deep ditches are better than broader, more shallow ones; fourth, that the cost is not heavy, and fifth, that it is quite feasible to eliminate the salt marsh mosquito entirely on the marsh areas that can be ditched."[20]

In early February Quayle met with Pope and outlined his proposal. Patterned on Smith's work at Rumson's Neck and Monmouth Beach in New Jersey, Quayle's plan called for improving drainage by digging a network of narrow, deep ditches and filling areas to prevent breeding. Oil would be used as a last resort when permanent measures could not be employed. The Club's executive committee approved the plan. After the meeting Pope took Quayle aside and told him "that the mosquito must go at any cost."[21]

Work began on February 27, 1905. The Improvement Club supplied $2,000 for the project. Eventually Quayle hoped to bring relief to a ten-mile stretch of salt marshes running from South San Francisco northward to San Mateo. He hoped that the Burlingame campaign would serve as a model for future projects. Following Smith's lead, Quayle and his students searched for the prime larval habitats. It would be prohibitively expensive to dike and ditch all of the salt marshes along San Francisco Bay. Moreover, not all of the salt marsh produced mosquitoes. By early February Quayle had identified the source of much of Burlingame's salt marsh mosquito. "I am fairly certain," he told Smith, "that the key to the whole situation is an area of 600 to 800 acres that is surrounded by a dike built some 10 or 15 years ago."[22] Quayle ordered a crew to repair the dike, fill cracks, and dig ditches in the surrounding marsh.

In June he reported to Smith that "the results are so far successful beyond any expectation that I ever had and I feel confident that they can be kept so for the rest of the season." The anti-mosquito work transformed Burlingame. A year earlier "gardeners and others working out of doors were obliged to wear netting for protection."[23] Nearby communities such as San Mateo, Hillsborough, and San Rafael benefited from the Burlingame work. Burlingame's success, in fact, led nearby San Mateo's trustees to enact California's first anti-mosquito law. In July the city's trustees declared any water that bred mosquitoes was a "public nuisance." The trustees authorized San Mateo's health board to issue fines and impose jail sentences on residents who failed to comply with City Ordinance 106 "forbidding the maintenance of all places breeding mosquitoes."[24]

Woodworth was delighted with Quayle's success. In May he won approval to expand the Station's anti-mosquito work. He wrote to Smith inquiring if he had "any bright young people available for an assistantship in entomology" at the California Station.[25] "I am sorry to say," Smith replied, "that there is nobody available. None of the boys coming on in this class have at all specialized in my line; and those who have applied to me recently are not such as I would care to recommend."[26]

When Woodworth's letter arrived in New Brunswick, Smith was in the midst of a bruising fight over the anti-mosquito movement's future. Governor Murphy's approval of the Duffield Amendment at the end of the previous year's legislative session had given local health boards the power to eliminate nuisance mosquitoes within their jurisdiction. Controlling salt marsh mosquitoes was the remaining great challenge. Smith hoped that his *Report on Mosquitoes* would furnish guidelines for future permanent control work in the Garden State. Smith was confident that the progressive forces in Trenton would support his call for remedying the salt marsh problem. He had not anticipated, however, the mischief of Henry Clay Weeks.

Born in 1846, Henry Clay Weeks had a earned reputation at the turn of the century of being a friend of shade trees, a tireless advocate of buried telephone and utility lines, and most recently, a self-proclaimed "mosquito engineer." Weeks conceived of "mosquito engineering" as the central element in a multimillion dollar crusade that would transform the salt marshes along the Atlantic Coast into valuable real estate. His first anti-mosquito work was near Oyster Bay on the north shore of Long Island. By 1903 he had completed work on Centre Island at Oyster Bay and started a new project on Coney Island.

Coney Island was notorious for its huge numbers of mosquitoes. Summer visitors complained of being "besieged by great swarms of mosquitoes." Horse racing fans and animal lovers commented on the "torture [mosquitoes] inflicted upon the blooded horses each year stabled at the Sheepshead Bay Race Track."[27] "The situation as a whole," Weeks declared in an interview *New York Times* reporter, "is one of grave magnitude. The conditions for breeding mosquitoes are perfect, and such as to earn for the section an unenviable reputation of a monumental character."[28]

Anti-mosquito work on Coney Island began early in 1903. During the previous fall Weeks secured the financial backing of the Long Island Railroad, the racetracks, and the New York City Health Department for his anti-mosquito plan. In May he organized a "mosquito luncheon" to publicize the project's accomplishments. More importantly, Weeks announced the formation of "a body of citizens, scientists, and officials [who] have banded themselves together as a sort of vigilance committee, with the avowed object of making war on the winged pest."[29]

Weeks recruited the members of his mosquito "vigilance committee" from New York's political and financial elite. William Whitney, a prominent New York attorney and a former secretary of the Navy, served as the committee's leader. Other members included the railway tycoon, J. Stanley Brown, J. F. Calderwood of the Brooklyn Rapid Transit Company, Cornelius Fellowes, the president of the National Horse Show and secretary of the Coney Island Jockey Club, and Harry Macdona, a former New City assistant district of attorney. At the luncheon's conclusion Ernst Lederle, president of the New York City Department of Health who later was criticized for his decision to release Mary Mallon, Typhoid Mary, from detention, pledged $10,000 to support the "mosquito crusade."[30]

Weeks envisioned a "mosquito crusade" that would eliminate mosquitoes through a combination of "filling and draining and petrolizing." Unlike Smith, Weeks believed that it was necessary to remove all water from the marshes. Weeks considered ditching that allowed water to flow in and out of salt marshes a half measure. Victory over *Aedes sollicitans* would be achieved when the marshes were dry.[31]

Over the next six months Weeks grew enamored with the idea of expanding his mosquito "vigilance committee." "The [anti-mosquito] movement has long passed the humorous stage," Weeks explained. Across the nation, progressively minded citizens were forming "mosquito brigades," improvement societies, and clubs. Weeks believed that there was a need for an organization that would "let people know what is being done, and by whom and the how and why of it."[32] In October he issued an invitation to New York's business elite and those interested in mosquito control to meet on December 16, 1903, at the Board of Trade and Transportation building on Broadway in New York City to discuss launching a national mosquito crusade.

More than one hundred business executives, politicians, lawyers, engineers, physicians, and entomologists attended the "First General Convention to Consider the Questions involved in Mosquito Extermination." Large placards greeted the participants as they entered the meeting room. Attendees were encouraged to "form the anti-mosquito habit" and "let our motto be: "No stagnant water." A huge poster quoting Leland Howard proclaimed the meeting's theme: "Mosquito extermination is not a temporary interest but the beginning of a great and intelligent crusade."[33]

The meeting lasted three hours. Speakers were limited to six minutes for their presentations. In all seventeen papers and letters were read. At the meeting's conclusion the attendees passed a resolution calling for the creation of the National Mosquito Extermination Society. The society would meet annually. They named Henry Clay Weeks the society's secretary and spokesperson.

Neither Howard nor Smith attended the meeting. At the last minute a summons to testify before a congressional committee forced Howard to cancel his presentation. Smith was at work in Camden. Both men, however, submitted papers, which Weeks published in the conference's Proceedings. In his prepared remarks Howard emphasized that the New York meeting was part of a "Worldwide Crusade." In other countries the campaign to eliminate mosquitoes had developed as a public health initiative. Eliminating mosquito-borne diseases was not the primary motivation of either the South Orange Improvement Association or Week's "vigilance committee." "The main incentive," Howard observed, "to all this world-wide movement has been the prevention of disease. Probably nowhere else in the world has the motive of personal comfort entered into the crusade as it has in the United Sates, and we have already carried this aspect of the work much further than any other country."[34]

In his paper Smith argued that sound entomological research must guide the movement. There were, in fact, multiple mosquito problems. His work in New Jersey illustrated this. During the previous two years, he had identified thirty-one species of mosquitoes in New Jersey. Of these, no more than seven or eight were troublesome to human beings. A handful of these mosquitoes posed a public health problem. These mosquitoes were the responsibility of the local boards of health. The great majority of the state's most troublesome mosquitoes posed no public health risk. They were "confined to salt marshes for breeding places—the adults unfortunately have a wider range." Careful observation demonstrated that two species of salt marsh mosquitoes had a migratory range of up to forty miles.[35]

New Jersey, Smith explained, faced two distinct mosquito problems: mosquitoes of limited flight range and salt marsh mosquitoes with long migratory patterns. The first were properly the responsibility of local communities. "The state," Smith declared, "should not go further than to furnish advice and, perhaps general direction for the work." A solution for the salt marsh problem where "millions of mosquitoes migrate to communities in other counties that have no jurisdiction over the breeding places . . . is as yet a little obscure." Solid entomological research needed to be done before launching any effort to solve the salt marsh problem. "In brief," Smith advised, "my idea is that we should first of all spend time and money learning the character and extent of our task—that we may save time and money wasted in makeshifts, and the loss of confidence resulting from failures."[36]

Weeks disagreed with Smith's cautious, sober approach. The public demanded bold action. It would be a mistake to expend more time on research. Accordingly, he ended the meeting with a call to arms, or more precisely, a call for engineers. "Mosquito extermination is essentially an engineering problem," Weeks declared. Engineers should lead the mosquito crusade. Certainly, there was a place for entomologists in the movement. Practical men of action, however, would resolve the challenges of the twentieth century. In the future, Weeks concluded, "we may all be mosquito engineers . . . and after a regular course of work we will be entitled to put M.E. [Mosquito Engineer] after our names."[37]

Weeks's National Mosquito Extermination Society attracted little attention during the next twelve months. A few newspapers carried brief reports of its formation of the conference. The editors of *Engineering News* applauded Weeks's call for "mosquito engineering." They published a detailed summary of the conference papers. The Society's executive council planned an expanded two-day meeting for December.

The Extermination Society's second annual meeting provided no hint of the storm that lay ahead. At the December 1904 meeting the delegates agreed to change the organization's name to the American Mosquito Extermination Society to reflect the Society's continental aspirations. Leland Howard, William Gorgas, Spencer Miller, and John Smith all contributed papers. Not one to miss an opportunity to attract publicity, Weeks persuaded Frank W. Moss, New York City's police commissioner, to open the first session. Moss announced that he was issuing a criminal indictment of "All Humanity versus All Mosquitoes." Moss charged that the "defendant mosquitoes, *Culex*, not having the fear of God or man before their eyes" having committed numerous crimes should be found guilty and receive "prompt execution."[38] Moss promised to use his powers as police commissioner to aid the society in taking immediate action against criminal mosquitoes.

The first hint of discord within the anti-mosquito movement came shortly after New Year's Day. Unbeknownst to Smith, Weeks had launched a two-pronged legislative initiative that threatened to undo Smith's work. In 1904 Weeks persuaded Townsend Scudder, who represented Long Island in Congress, to introduce a bill calling for the creation of a federal mosquito commission. The bill specifically charged the twelve-member commission with oversight of the New Jersey and New York salt marsh problem. Simultaneously, Weeks asked Godfrey Mattheus, an assembly member from Jersey City, to introduce Assembly Bill #250, which provided for the creation of a mosquito commission to supervise anti-mosquito work on the Hackensack meadow and along the Passaic River.

Smith was shocked when he learned of Weeks's plan. He believed that Weeks's proposal for a federal mosquito commission jeopardized his effort to expand his anti-mosquito demonstration projects along the Jersey Shore. He feared that the New Jersey legislature would use Weeks's federal proposal as an excuse to suspend their support for mosquito control. In early January Smith persuaded James Hildreth, a Cape May assembly member, to introduce a bill providing state support for local health boards in their fight against mosquitoes. The Duffield Amendment had given health boards the power to declare waters containing larvae a public nuisance. The amendment, however, left open the question of precisely how local boards were to pay for the work. Hildreth's bill granted state relief to communities undertaking anti-mosquito work on a matching basis.

Initially there was wide support for Smith and Hildreth's bill. Real estate developers believed that it would increase property values and sales. The members of the New Jersey Conference on Mosquito Extermination thought it a first step in easing the burden of local health boards. Smith feared that Edward Stokes, the new governor, might block the mosquito movement advance. Stokes, he learned from Clinton Sellers, director of the Cape May Real Estate Company, was of two minds about mosquito control. Sellers sent Smith a copy of a letter that Stokes had written to Tom Millett, mayor of Cape May City. "I . . . am thoroughly convinced," Stokes wrote, "of the fact that Mr. Smith has made a great success of his plan for the extermination of the mosquito. I have not however examined Mr. Hildreth's bill for the appropriation of twenty thousand dollars. . . . Generally speaking I am opposed to adding any more charges to the State's already heavy burdens."[39]

In early March the New Jersey Assembly and Senate passed Hildreth's bill. Stokes, as Smith predicted, "sat down hard on the Mosquito Bill."[40] Rather than using his veto, Stokes allowed the bill to languish on his desk. Believing there might be one final opportunity to win Stokes's support for an amended version of the bill, Hildreth and Smith redrafted the proposal. The amended version significantly lowered the amount of funds the state would provide in support of the local health boards.

Weeks's effort to create both a national and a New Jersey mosquito commission added to Smith's difficulties. Smith contended that Weeks's proposal gave Governor Stokes an excuse to "sit on the [Hildreth] bill" and wait to see what Congress would do. Weeks's arguments revealed that he knew little about entomology and less about New Jersey's mosquitoes. "You will see," Smith wrote to Benjamin Howell, his congressional representative, "that this whole subject has been fully considered in our State." The *Report on Mosquitoes* demonstrated, in fact, that there was "very little Hackensack meadow problem."

Smith dismissed Weeks's claim to represent a national constituency. "This National Mosquito Society, for which Mr. Weeks apparently speaks, consists mainly of one individual with a lot of persons he has associated with him in most cases without consulting with them and without their consent."[41]

More was at stake in the Weeks–Smith dispute than the formation of a state and national mosquito commission. Weeks's earlier work near Oyster Bay on Long Island revealed his real objective. "Piece by piece marsh and meadow land will be taken up," Cromwell Childe wrote in his summary of Weeks's 1902 work, "and made mosquitoless, drained, and dyked. It is costly work. . . . but the association does not fear the expense. Too many wealthy men and men of progressive ideas are within these bounds. All this spells improvement, the adding of large values to property very plainly." Weeks hoped to use a national mosquito commission to rally support for the "new science . . . of Mosquito Engineering."[42]

Smith disagreed with both Weeks's methods and his objective. His study of New Jersey's salt marshes demonstrated that not all of the marshes produced mosquitoes. Moreover, it was unnecessary to eliminate the marshes. His experiments at Rumson Neck and on the Newark meadow demonstrated that a network of narrow, deep ditches allowed excess water leave the marsh while allowing predacious fish access to inland pools. Smith thought that Weeks's proposal had more to do with making money than killing mosquitoes. "The only real problem," Smith told Howell, "is Mr. Weeks's problem of how to get employment as an engineer on some extensive piece of work with a government appropriation behind it."[43]

By mid-March Smith was certain he had blocked the Weeks's effort. "I want to kill [Assembly Bill #250] and I think the death blow has been delivered," Smith wrote to Henry Kummel, New Jersey's state geologist. "But if we can kill it just a little deader than it is already, by convincing Mr. Mattheus [who introduced Weeks's proposal to the Assembly] that we already know something about the Hackensack Marshes no harm will be done."[44] Smith arranged to meet with Godfrey Mattheus. He concluded that the assembly member "was a very nice sort of fellow" who had been duped by Weeks.[45] After the meeting Smith sent Mattheus a point-by-point refutation of each of Weeks's claims. "I think you will realize," Smith concluded, when you look over the ground, that Mr. Weeks has simply used you to further his own ends and without giving you the information you should have had before introducing a measure like Assembly #250."[46]

Confident that he "killed" Weeks's bill "deader," Smith rallied support for the amended version of Hildreth's bill. Smith penned a handwritten appeal to his old ally Edward Duffield asking him to lead the fight for the Hildreth's amended mosquito bill. "The failure of this bill," Smith wrote, "would put back the mosquito extermination movement in this state at least five years and I know you don't want to contemplate that."[47] The redrafted version of Hildreth's bill passed both chambers of the legislature at the end of March. Smith immediately went to Trenton and secured a private meeting with Governor Stokes. "I saw the Governor," Smith told Hermann Brehme. "He promised me positively that as soon as it comes to him he would sign it."[48]

Smith waited until the end of the legislative session to vent his fury on Weeks and the American Mosquito Extermination Society. "You [have] showed," he declared, "your complete ignorance of New Jersey affairs." He charged Weeks with misrepresenting the *Report on Mosquitoes* and the state's exhaustive geological survey of the Hackensack and Passaic valleys. Weeks had ignored the conclusions of "engineers who really understood their business" in his effort to make secure work for himself. "I am decidedly interested in discovering just how the business of the American Mosquito Extermination Society is carried on and whether it is part of its settled policy to act in the underhanded manner pursued in New Jersey." As far as Smith was concerned, the American Mosquito Extermination Society was more "a hindrance instead of an aid to rational methods of dealing with the mosquito question in New Jersey at least."[49] Weeks should leave the matter of mosquito control in New Jersey and New York to those suited to the task.

While Smith was fighting back Weeks's foray into New Jersey, he found time to encourage the development of "rational methods of dealing with the mosquito question" in New York and Delaware. Alvah Doty provided Smith with an unexpected opportunity to demonstrate the effectiveness his ditching program in New York. Since his 1901 anti-malaria campaign on Staten Island, Doty had continued to rely exclusively on temporary means of mosquito control. Early in May 1905 the New York newspapers reported that Doty planned to expand his control effort. Smith discounted the effort because he was convinced that Doty would waste the money on oiling. "The whole matter has been referred to Dr. Doty," he told Hermann Brehme. "The net result of his work will be, one thousand dollars spent, and nothing accomplished. I am heartily sorry that this should have been the outcome."[50] Despite his misgivings Smith wrote offering his assistance and the hope that "you would find it possible to give an object lesson in dealing with the permanent improvement of salt-marsh areas by drainage. . . . New Jersey," he concluded, "is directly interested in the salt marsh areas along the west shore of Staten Island. . . . In a series of captures made at South Orange several days ago, *sollicitans* was found, and absolutely the only place the species could have come from is the Staten Island brood."[51]

Three days later he received an urgent telegram from Doty.[52] Doty was not only willing to consider instituting a permanent control plan on Staten Island but he also wanted Smith's help in designing and implementing it. The next day in a phone conversation Doty and Smith agreed to meet on Staten Island. Doty invited the borough's president and several members of the city's health board to accompany him. Smith brought Eugene Winship and Hermann Brehme. By the end of the morning the two groups had reached an accord. Under Smith's supervision Winship and Brehme would prepare a survey of Staten Island and develop a drainage plan. The health board and the borough of Richmond agreed to finance the project. Smith's one stipulation reflected his political sensitivity. Smith did not want to offend Porter Felt, New York's state entomologist. Thus he insisted that his name "not appear in the matter in any official capacity, because New York already had an official who worked along the same lines, and was fully competent to do the work."[53]

Smith made a hurried trip in mid June to Delaware while Winship and Brehme completed their cost estimate for the Staten Island work. The meeting was an important one. Smith had agreed to describe New Jersey's mosquito work to Governor Preston Lea, state officers, the chancellor of the university, and members of the state grange. In May Smith's attendance at Rutgers' commencement compelled him to reschedule the meeting to June. "I only go to church once a year," Smith explained, "and that is on the day of the Baccalaureate sermon; and if I should miss that important event, people would begin to look upon me as irreligious."[54]

The talk was a success. "Referring to our conversation on Saturday . . . concerning the mosquito question," Smith wrote to Governor Lea, " . . . something of this character could readily be done for the State of Delaware through the experiment station, and without any very great expenditure of money."[55]

Smith's work with Doty on Staten Island and his efforts in Delaware underscore his differences with Henry Clay Weeks. Unlike Weeks, who believed in a single brand of "mosquito engineering," Smith's work in New Jersey convinced him that effective control strategies developed out of a close study of local conditions. Smith did not want Delaware to adopt an anti-mosquito program that mimicked New Jersey. Rather he encouraged Delaware's leaders "to investigate in Delaware conditions along the lines followed in New Jersey."[56] Smith was adamant. Entomology and ecology must guide anti-mosquito control work.

Winship and Brehme's efforts on Staten Island reflected this principle. They prepared a detailed report that identified the principal larval habitats along the western shore. Winship estimated that the total project would cost no more than seventeen thousand dollars. Smith told Doty that he considered this a reasonable figure given the extensive amount of work. It was still a substantial sum. Both men breathed a sigh of relief when they learned that George Cromwell, president of the Borough of Richmond, supported the proposal. "I am more than pleased," Cromwell told Smith, "at the low figure for which you think the drainage of all the marsh areas on Staten Island can be effected. I have laid the matter before the Mayor of the city and he has promised to take the trip with Dr. Darlington, Dr. Doty and myself tomorrow to inspect marsh lands in person."[57] The mayor and the board of estimate approved the project. Winship and Brehme immediately set to work under Smith's unofficial direction. "Whether matters can be carried through as rapidly as Dr. Doty seems to hope," Smith wrote to Porter Felt, "of course I do not know."[58] When completed, Smith was confident the residents of Staten Island as well as nearby Elizabeth and Newark would experience an increase in both comfort and property value.

Matters of comfort were not the principal concern of health officials in upstate New York, Louisiana, and Florida in 1905. Typhoid and malaria epidemics swept through Ithaca, New York, in 1903 and 1904, prompting city officials to institute a citywide sanitary and anti-mosquito campaign. At the beginning of the twentieth century Ithaca had a reputation of being "a very dirty town, full of ancient evils."[59] The town's location near Lake Cayuga in the basin of a "low sodden plain" added substantially to its unhealthy character.

Ithaca's unsavory sanitary conditions prompted the city and Cornell University officials to launch an anti-mosquito campaign the following year. The situation reached a crisis point in the summer of 1904. Malaria swept through the city. By December local physicians had documented more than 2,000 cases out of a population of 10,000 within the city.[60]

Concern about the mosquito menace surfaced during the 1903 typhoid epidemic. Dr. George A. Soper, a local physician, observed anopheline mosquitoes throughout the city and warned of a possible malaria outbreak. In July 1903 Ithaca's board of health declared that the "malaria situation was serious" and recommended the city take immediate action. Fearing a loss of students, Cornell's president offered to underwrite one-third of the cost of the mosquito eradication work.[61] In February 1905 Ithaca's mayor sent an urgent request to Albany requesting the legislature's support for the town's anti-mosquito initiative. New York State's first anti-mosquito legislation gave cities the power to "take over, condemn, in any other manner acquire all marshland within the city limits, to supervise and investigate all mosquito-breeding areas on private property, or in any other way control the breeding of mosquitoes."[62]

A year elapsed before the legislature approved the bill permitting the formation of an anti-mosquito commission. City and university officials did not wait for Albany's approval. On April 17, 1905, the city council approved a local mosquito control ordinance based on Tompkins County Medical Society's anti-mosquito recommendations. This decision earned Ithaca the distinction of being the first American city to institute a malaria campaign based on mosquito control. Mosquito control work began in June. During the height of the malaria season, city mosquito inspectors applied oil to all standing bodies of water within the city every ten days. By 1912 the malaria threat receded to a point that the city discontinued its anti-mosquito measures.[63]

While Ithaca's health inspectors struggled to eliminate anopheline larvae, rumors of "Yellow Jack's" return to the Crescent City spread along the Gulf Coast in July 1905. The Reed Commission did not resolve the debate about yellow fever. Many physicians and local health board discounted the commission's conclusions. Cities such as New Orleans had invested considerable sums of money in developing and maintaining elaborate quarantine procedures. The Louisiana Health Board "acknowledged the mosquito as one factor in the conveyance of yellow fever" in its 1900–1901 annual report. The state's health authorities, however, clung to the notion that the yellow fever spread through fomites (tiny particles) that could be absorbed in clothing. "We Southern Health Officers," a Louisiana official declared, "charged with the grave duty of protecting our people against this most dreaded of diseases are unwilling to accept the dictum of experimenters that yellow fever can be conveyed by no other agency" than the mosquito.[64]

Dr. Quitman Kohnke led the campaign in New Orleans to introduce anti-mosquito measures. Reports of Gorgas's success in Havana convinced Kohnke, who served as head of the New Orleans's board of health in 1901, to begin a citywide anti-mosquito campaign. The public, however, showed little interest in his drive to screen houses and cover cisterns. Local political leaders rejected Kohnke's recommendation that New Orleans adopt an anti-mosquito ordinance.

These setbacks did not diminish Kohnke's enthusiasm for mosquito control. Kohnke traveled to New York in 1904 to take part in the second meeting of Weeks's mosquito extermination society. Kohnke told the delegates that he planned to renew his anti-mosquito campaign in 1905.[65]

At first apathy contributed to Kohnke's failure to win public support for mosquito control. There were no cases of yellow fever in New Orleans between 1900 and 1904. Many of the Crescent City's 375,000 residents had never experienced a yellow fever epidemic. This changed on July 13, 1905, when a New Orleans physician noted a handful of suspicious cases in a city ward populated largely by Italian immigrants. Health officials waited a week before admitting yellow fever was present in the city. Two days later Kohnke and Dr. Beverly Warner, rector of Trinity Episcopal Church, launched an anti-mosquito campaign.

In early August New Orleans mayor Martin Behrman met with Kohnke, members of the state Board of Health, and prominent business leaders to map out a strategy for the city. At the meeting's conclusion Behrman issued a request for federal intervention. President Theodore Roosevelt approved the request and ordered Surgeon General Walter Wyman and the U.S. Public Health and Marine Hospital Service to take charge of the anti-yellow fever campaign.[66]

The federal assumption of responsibility for the epidemic sparked protests. The New Orleans *Picayune* criticized Behrman's decision. "Now we rush into the arms of Uncle Sam," the *Picayune*'s editor declared, "and are only too happy to trade our out-of-date Democratic State sovereignty . . . for a temporary sanitation of the city."[67] The success of the anti-mosquito campaign assuaged the hurt feelings. By October the epidemic was in retreat. At the end of October Theodore Roosevelt visited New Orleans. At a banquet in Roosevelt's honor, Mayor Behrman thanked the President for his support of the city. Roosevelt praised the courage of New Orleans in its fight against the mosquito pest and declared "that at any moment, if he had been asked to do so, he would have come in person to assist in this fight that was being so gallantly made."[68]

The 1905 New Orleans yellow fever epidemic sent shock waves across the south. Joseph Porter, head of Florida's state health board, had suspected that New Orleans officials, fearing adverse publicity, had suppressed knowledge of the outbreak. Porter sent a coded telegram to Surgeon General Walter Wyman requesting details on the outbreak. Wyman's reply arrived too late. Five hundred people had already returned to Pensacola from a three-day "cheap excursion" to New Orleans. They brought the epidemic to Florida's Gulf Coast.[69]

Porter was a vigorous advocate of the anti-mosquito movement in Florida. His concern about the potential threat of mosquitoes led him to assign his assistant, Hiram Byrd, the task of surveying the state's mosquitoes. Byrd's *Mosquitoes of Florida* appeared in 1905, six weeks before the outbreak of yellow fever in New Orleans and Pensacola.

The forty-seven-page *Mosquitoes of Florida* was a shadow compared to Smith's voluminous *Mosquitoes of New Jersey*. Nevertheless, Byrd shared Smith's commitment to science. Most of *Mosquitoes of Florida* is devoted to careful descriptions of the life histories of the twenty-two different species of mosquitoes Byrd found in Florida. Byrd followed Smith's lead in emphasizing the significance of the migratory range of salt marsh mosquitoes. He assigned two salt marsh species, *Aedes taeniorhynchus* and *Aedes sollicitans*, with responsibility for most of the discomfort and economic injury suffered along Florida's Atlantic and Gulf coasts.

Byrd closed the *Mosquitoes of Florida* on a prophetic note. Like Smith, he envisioned a future in which entomologists, engineers, and public health workers would rid Florida of the mosquito blight. "The time is coming," he observed, "and not far distant, when the toleration of mosquitoes will be a municipal crime." Mosquito control was not a utopian dream. "That [mosquitoes] can be controlled is no longer the question," Byrd concluded, "but a demonstrated fact and it now remains with us to decide whether we will be leaders or followers in this beneficent crusade."[70] The 1905 yellow fever epidemic in New Orleans and Pensacola indicated that Bryd's vision of a mosquito-less Florida lay in the future.

Yellow fever regularly laid waste to New Orleans in the nineteenth century. Between 1817, the first recorded outbreak of yellow fever in New Orleans, and 1905 more than 41,000 people had died of yellow fever in the city. The success of the anti-mosquito measures in limiting the 1905 epidemic was apparent. During the 1905 four-month-long New Orleans epidemic, there were only 437 fatalities. The 1905 yellow fever epidemics in New Orleans and Pensacola were the last outbreaks of the disease in the United States.[71]

While federal and state health workers sought to relieve New Orleans and Pensacola from the scourge of yellow fever, the battle between Smith and Weeks had resumed. In April 1905 Smith wrote to each member of the American Mosquito Extermination Society's executive committee. He declared that he wanted to know if this "Society acts through a responsible committee or through an irresponsible Secretary who seems to have assumed the right of speaking for it." Smith objected to the inclusion of his name on the Society's letterhead as a member of the advisory board of entomologists. This was particularly galling because he had not given his permission. Worse, the society's introduction of Assembly Bill #250 was a "distinct discourtesy to and reflection on" Smith's work. "Is it the policy," Smith asked, "of the Society to ignore and condemn as unworthy of consideration all work not done under its seal?"[72]

The responses Smith received from the members of the American Extermination Society's executive committee confirmed his suspicions. "I regret to say that the press of other duties has made it impossible for me to attend the meetings," Paul Cravath, an influential Manhattan attorney and committee member replied. "It therefore happens that I have no personal knowledge of the matters to which your letter refers."[73] William Matheson, the committee's chairman and the society's president, answered in a similar vein. "It is quite impossible," he wrote in justification of the committee's lack of oversight to the society, "for men who have outside responsibilities that you will recognize that many of the officers and committeemen of this Society have, to attend to all the minute details which have, in our case, been taken up chiefly by the Secretary." Matheson assured Smith that he was confident that it was a mere "oversight" on Weeks's part that Smith's name had "been printed without your authority." Smith, Matheson claimed, had misconstrued the society's motive behind Assembly Bill #250. "Nothing could be further from the intention of the Society or its Executive Committee than to in any way interfere or compete . . . certainly not with anyone who has done the splendid work and attained the eminence in this field which you have."[74]

Nearly two months passed before Smith received Weeks's feeble apology. Weeks claimed that illness brought on by "over work in our cause" had prevented his writing. Smith had misunderstood his motives. "There is nothing," Weeks claimed, "I think in it that now requires answering." In the future, Weeks declared, he would "be much obliged" if Smith would inform him when the American Mosquito Extermination Society presented a hindrance. "If one cannot aid," Weeks concluded, "one certainly do [*sic*] not wish to hinder."[75] Weeks assured Smith that there would be no future disagreements.

Four months later Smith was shocked when he opened the October issue of *Entomological News* and once again found himself listed as one of the American Mosquito Extermination Society's expert advisors. "The public advocacy of scientific relief from mosquitoes and communities is rapidly growing," Weeks explained in an accompanying article. The society's efforts to encourage this growth had already led "to many crusades." Smith's astonishment must surely have turned to anger when he read further and learned that Weeks claimed credit for the recent work on Staten Island. "A general body like this can, without just cause of offense, strive for the cure of monumental evils by legislation and otherwise, as it has done in the case of the New Jersey–Staten Island marshes, where only great engineering plans, involving many interests, will permanently change conditions," Weeks explained.[76]

Weeks underestimated Smith. Smith's blistering response appeared in the next issue of *Entomological News*. "I will ask permission," Smith wrote, "to explain that while the professions of the Society are sound its actions through its Secretary are underhanded, dishonest, and not in accord with recent developments as to practice." Smith declared that he had never given Weeks permission to use his name. Weeks claim that he had contributed to the Staten Island and New Jersey mosquito work was bogus.[77]

Smith's letter chronicled Weeks's dishonesty. Smith was still angry at Weeks's attempt "to sneak through a bit of legislation" in New Jersey in the spring. If Weeks had been acquainted with the matter, he would have known that there was no need for a mosquito commission to oversee the work on the Hackensack meadows. "Fortunately," Smith observed, "the ignorance behind the bill was equal to its dishonesty . . . and the bill died the day I learned of its introduction."[78]

The Rutgers entomologist saved his full fury for Weeks's brazen references to the New Jersey–Staten Island problem. "Neither Mr. Weeks nor the Society for which he speaks," Smith observed, "has been consulted, and, therefore, may not know that so far as New Jersey is concerned more than half the work involved in the New Jersey–Staten Island problem is already done; that so far as Staten Island is concerned a survey of the territory was made under my direction and a report made that $17,000 would do all the necessary work." By the fall of 1905 the first phase of the Staten Island ditching was nearly complete. Weeks was an imposter. "It is always unpleasant," Smith concluded, "to introduce a personal element in a matter of public concern, but the publication of the circular in the [*Entomological*] *News* leaves me no option . . . the American Mosquito Extermination Society is, in my opinion, simply an attempt to advertise its Secretary, and not an effort directed to the extermination of the mosquito pest."[79]

Smith allowed his anger to get the best of him. The claim that half of New Jersey's salt marsh work was completed was grossly overstated. In the midst of his fight with Weeks, Smith confided to Spencer Miller that anti-mosquito ditching along the Jersey Shore was just beginning. Smith had completed his survey of the New Jersey salt marshes by December. It had cost $17,000 for the previous summer's ditching on Staten Island. Extrapolating from these figures, Smith estimated that it would require at least $300,000 to address the Garden State's salt marsh mosquito problem. He asked Miller if he would use his influence with his South Orange progressive friends "something of this kind [an appropriation for $300,000 for salt-marsh ditching] to the legislature and see how they feel about it." It was, Smith concluded, essential to "familiarize the public with the matter." The future of mosquito control depended on the public's understanding and support.[80]

Smith's enthusiasm for state support for mosquito control in the salt marshes grew in the next three weeks. It answered the principal objections of both property owners and local governments to undertaking salt marsh mosquito work. As matters stood, the cost of ditching a property often exceeded the land's value. Cities faced a different challenge. The broods of mosquitoes that periodically persecuted their residents often originated outside their municipal boundaries. His logic was simple: If the benefit of winning relief from salt marsh mosquitoes would come to all of New Jersey, then all municipalities should share the cost.

In early January 1906 Smith began his campaign to rally support for the idea. He assumed that the New Jersey Conference Committee on Mosquito Extermination would support the salt marsh initiative. Formed in 1903 by Smith, the conference committee had become an independent, potent political voice. He was surprised when he learned "the Conference Committee had given only a perfunctory and half-hearted support to the measure" at its mid-January meeting. If true, Smith observed, "it will probably result in a failure to act upon the measure that is ready for presentation to the Legislature."[81]

Fortunately, Smith had powerful friends in the New Jersey Senate who were ready to back his proposal. After his conversation with Spencer Miller, Smith had prepared a draft bill and persuaded Oliver H. Brown, a senator from Monmouth County, to introduce it as Senate Bill #82. Brown, a banker and real estate investor who served as Spring Lake's mayor for thirty-two years, believed that mosquito control was essential for seaside community's development.[82] Brown's bill, Senate Bill #82, called for the state to authorize up to $350,000 for salt marsh work. The bill placed a $70,000 ceiling on the expenditures in any single year. The director of the state experiment station or his executive officer, the state entomologist, was responsible for approving the plans and budgets for all work on the marshes. Senate Bill #82 gave John B. Smith the power "to start work systematically and to arrange matters in such a way that communities that are really anxious to do things for themselves will be helped to the extent of the State's power, and that pressure will be brought to bear wherever possible in those areas where the work should be done."[83]

Smith's preparations for a trip to Europe added to the feverish pace during the closing weeks of January 1906. On January 27 he planned to depart for a two-month visit to Rome and Naples. He worried that his absence during the legislative session would harm the bill's prospects. In the week before sailing Smith prepared detailed instructions for Augusta Meske, his sister-in-law and stenographer. During his absence, Meske would be responsible for overseeing his Rutgers laboratory, answering his correspondence, and sending him weekly summaries. Spencer Miller, Hermann Brehme, and Eugene Winship promised to go to Trenton and testify before the relevant Assembly and Senate committees. Edward Voorhees, the experiment station's director, assured Smith that he would do everything in his power to secure the bill's passage.

On the eve of his departure, *Entomological News* published Weeks's biting rebuttal to Smith's November letter. No one escaped Weeks's criticism. The editors of *Entomological News* were remiss in publishing Smith's "injurious letter" without having made "some inquiry as to whether there might not be an animus beneath and no basis for such serious statements." Responsible editors of a scientific publication, Weeks declared, should have been "cautious in uttering a libel."[84]

Weeks portrayed himself as an innocent victim. He claimed that he first learned of Smith's "bitterness" in April. Weeks acknowledged that Smith had written "so far as the list [entomological advisors for the American Mosquito Extermination Society] is concerned, my name appears without my authority." It was, Weeks maintained, untrue that he had used Smith's name "without permission." He claimed to have immediately removed Smith from the Society's roster upon receipt of Smith's August letter. The October circular that had included Smith's name was an oversight.

Weeks alleged that Smith had opposed his proposal for a state mosquito commission because of wounded pride. "The Trenton bill," Weeks observed, "mentioned certain officials for its commission." Smith resented that his name was not included. Weeks maintained that Smith opposed the legislation because this "bill was not one for entomologists but mainly for engineers." When Smith realized this "he worked himself into a rage, dangerous to himself and everyone else" and vented his bile in "abusive letters." Smith's real complaint, Weeks asserted, was that the American Mosquito Extermination Society represented an independent, national movement.[85]

Weeks ended his harangue in a manner calculated to infuriate Smith. Weeks portrayed himself as the injured party. He had devoted himself to the mosquito crusade. Smith's villainy grew out of a "strange jealousy" for Weeks's selfless work and his accomplishments. "May I assure your readers of the fact," Weeks declared, "that I have given some of the best years of my life to the society and cause and broken my health largely through these efforts." Smith's penchant for drawing "positive generalizations on insufficient data" compelled Weeks's response. Weeks would have preferred to "await results." The publication of Smith's calumnies had forced Weeks to respond. "Usually a person of the mind [John B. Smith] seems to have," Weeks declared, "is only encouraged by the weakness of his victims." In short, John B. Smith was a bully.[86]

In the hours before boarding the steamship *Princess Irene* in New York City, Smith penned a cheery six-sentence reply to Weeks's diatribe. "Just a word," Smith wrote, "to acknowledge a few minutes of real enjoyment in reading the letter of Mr. Henry Clay Weeks in the January *News*." Weeks had not refuted any of his statements. The *News*' readers could verify the facts. "Its very violence makes it unnecessary for me to reply," Smith observed, "but I do wish to disclaim any feeling of jealousy. So far as I am aware, Mr. Weeks has never done anything that any one need be jealous of. That he has done and is doing work in New Jersey may be true, because even in New Jersey there are men with more money than brains."[87]

After Smith's departure, a registered letter arrived in New Brunswick in which Weeks threatened, "to place the matter in the hands of my attorneys" unless a "full and complete retraction of, and apology for, the injurious and libelous statements thus made by you against me and my work."[88] Augusta

Meske forwarded the letter to Smith in Naples after scribbling, "the man certainly is off his base."[89] Smith heard nothing more from Henry Clay Weeks.[90]

During his two-month absence, Augusta Meske answered numerous requests for Smith's assistance in establishing anti-mosquito programs. "I regard the work to be undertaken in a large scale in New Jersey as of far more than local importance," Horace McFarland, president of the progressive American Civic League declared. "It means much to New Jersey, but it means more to the United States."[91] A year earlier McFarland, who was on the editorial board of the *Ladies' Home Journal*, had vowed "to do something to interest its very wide circle of readers in fighting mosquitoes." His goal was to use Smith's work to build a national movement. "It is," he explained, "thoroughly practicable to start a million people fighting mosquitoes, and that is what I am after. Your very competent help will be very highly appreciated."[92]

William Newell, secretary of the Crop Pest Commission in Louisiana, had concerns that were more pressing when he asked Smith to send a copy of *The Mosquitoes of New Jersey* to Robert Campbell. The 1905 New Orleans yellow fever outbreak generated new interest in mosquito control. In 1901 most physicians had opposed Quitman Kohnke's call for covering cisterns and oiling.[93] The 1905 epidemic encouraged sanitarians throughout the state. *The Shreveport Times* applauded Robert Campbell's selection to attend the state health conference following the New Orleans yellow fever epidemic. Campbell, a physician and leader of the local Progressive League, relied on Smith's work "with sanitation and the mosquito." An editorial writer for *The Shreveport Times* declared, "Dr. Campbell is thoroughly alive to the mosquito doctrine of transmission of yellow fever . . . and his ideas to the best methods of dealing with matters pertaining to public health are intensely practical."[94]

Support for Smith's own practical initiative, the salt marsh bill, grew in February 1906. The Newark, Kearney, and Orange health boards passed resolutions calling for approval of Senate Bill #82. Peter Shields, president of the Cape May Real Estate Company wrote to Senator Brown, "We wish to go on record as saying that the $350,000 that you have asked the State, will come back ten fold in one year, because, in our opinion, this is a business proposition with the State of New Jersey. Thousands of people have been driven away from New Jersey resorts and are staying away until they have some assurance that something has been done to eliminate these pests."[95]

The Senate hearing on the bill took place on March 13. Edward Voorhees, Spencer Miller, Eugene Winship, and Hermann Brehme testified. On its first two readings, the committee approved the bill only to vote it down on the third reading. Voorhees told Meske he was "working like the devil for the bill."[96] He promised to return to Trenton and win the committee's approval.

A week later the senate committee reversed itself and reported the salt marsh mosquito bill to the full senate. The Senate accepted the committee's recommendation and passed the bill.

Smith returned to America on April 3. He wrote to Thomas N. Gray, an East Orange physician and president of the Conference Committee on Mosquito Extermination, "I got back again to New Jersey just in time for a day's work in the Legislature. It was needed because the speaker had about made up his mind that he would not call up the bill until he was forced to do so."[97] Speaker Samuel Robbins, a Mount Holly Republican, hoped to postpone a vote on the bill until it was too late for the Assembly to act on it. Smith rallied support for the bill and forced a vote. The Assembly and the Senate voted to send the bill to Governor Stokes.

Convincing Stokes to sign the bill was not going to be easy. Louis Richards, the conference committee's secretary, and Spencer Miller suggested organizing a letter-writing campaign. Smith disagreed. "The governor is a peculiar sort of man," he explained, "and one who is not easily influenced by what may be called clamor." Smith was certain the governor agreed that mosquito control was beneficial. His problem with the salt marsh bill turned on a question of principle. The stumbling block was "why the State should do it rather than local communities. No argument which is not based so as to reach this particular point of view will have the slightest influence on him."[98]

Smith met with Stokes on April 16. He had asked A. L. Thornell, a member of the Monmouth Beach Improvement Association, and Alvah Doty to write to Stokes. Stokes knew these men and respected their opinion. During their meeting Stokes asked for details on the Staten Island project. More questions followed. Smith had an inspiration. If Stokes was willing, Smith would arrange an impromptu visit to Staten Island so the governor could see for himself what mosquito control entailed. Stokes accepted the invitation, and his secretary scheduled the visit for Saturday, April 19.

Smith had seventy-two hours to organize the governor's inspection trip. "This is our chance now to show what has been and what can be done," Smith wrote to Louis Richards immediately after his meeting with the governor. "I feel sure," he continued, "if we can make a respectable showing the governor will not only sign the bill; but help us in future [sic] to get necessary appropriations."[99] Smith met with the Gray and Richards on the Thursday before the Saturday excursion to plan the outing. Conference committee leaders would escort Stokes from Elizabeth to Staten Island. Doty and Eugene Winship would meet the governor's party at Howland's Hook. After disembarking Winship would show Stokes the difference between unimproved and improved sections of the marsh. It was imperative that Winship find a section of the unimproved marsh where there were "plenty of wrigglers." "He wants to see wrigglers," Smith told Thomas Gray, "as he has never seen any, so we must show him some."[100]

On Friday night as he was sitting down to dinner, Smith received a call from Stokes's secretary cancelling the excursion. One of Stokes's close friends had died earlier in the day during an appendectomy. During the phone call the secretary told Smith that Stokes had reviewed the salt marsh bill. "If the references made to existing laws as passed proved to be correct, the governor would sign it," the secretary told Smith. The secretary reviewed the questioned passages and confirmed their accuracy. The governor had signed the bill moments before leaving Trenton to arrange his friend's funeral. "I feel very well satisfied at the outcome," Smith told Doty, "because it commits the State finally to the work of extermination."[101]

Substantial challenges lay ahead. Money was a perennial problem. After Stokes signed the bill into law, the appropriations committee slashed the funding from $70,000 to $10,000 for the first year's work. "The legislature of the State of New Jersey," Smith later observed, "keeps me on pretty short allowance for mosquito work. It doles out its favors in small fragments and I never know until after the Legislature has adjourned exactly how much I have to spend."[102] Moreover, the law did not take effect until the following November. The "unfortunate mosquito man," as Smith had taken to call himself, would have to make do with what he had. There was no use to complain. "Now we are fixed in a general sort of way," he told Brehme, "and will have to show practically what we can do."[103]

Years of overwork were taking their toll on Smith. His family hoped that the trip to Europe would refresh him. Whatever energy Smith had regained was lost in his feverish drive to win passage of the salt marsh bill. Smith suffered an acute attack of nephritis in late April. "Nephritis as you know," Smith wrote to Alvah Doty, "is nothing to be fooled with. I am coming out of the acute condition fairly well, and hope to get about more actively in the near future. When I do, my first visit will be paid to Staten Island."[104]

From his sickbed, Smith was already mapping out the next season's work. The ditching in the Newark marshes needed to be completed. Later he planned to begin ditching along the Raritan River and in Cape May. Money, much more money, was required for the salt marsh work. The principal obstacle to ridding New Jersey from bloodthirsty, six-legged pests was two-legged politicians in Trenton. "I feel inclined to shout," Smith later wrote defining his outlook toward Trenton, "partly in the vernacular and partly in a sort of semi-vernacular go-it '*soc et tuum*.'"[105]

Chapter 3 A Continental Crusade

AT 5:13 A.M. ON April 18, 1906, the Great San Francisco Earthquake sent shockwaves that stretched five hundred miles from Coos Bay, Oregon, to Los Angeles, California. Many died instantly in the rubble beneath collapsed tenements. Gas lines ruptured throughout the city. The ensuing Great Fire raged for four days. Hundreds died in the earthquake and in conflagration that followed. Property damages were tremendous. Ineptitude and corruption magnified the losses. Progressives opened a statewide drive for reform. They demanded stricter building, sanitation, and fire codes. Hiram Johnson, a tough-minded Republican lawyer in San Francisco, directed his anger at corrupt politicians and profiteers. Johnson's four-year anti-graft campaign and efforts to regulate the Southern Pacific Railroad culminated in his 1910 victory in the Golden State's gubernatorial race.[1]

The San Francisco Earthquake and the Great Fire had significant ramifications for California's embryonic mosquito crusade. One year to the day before the earthquake, Henry Quayle of Berkeley sent an enthusiastic update on the Burlingame salt marsh mosquito work to John B. Smith. He was confident that the anti-mosquito movement would succeed. But after the earthquake, local support for mosquito control dwindled. The promising salt marsh work in Burlingame languished. Quayle's decision to leave Berkeley to accept a position with the Southern California Plant Pathology Laboratory in Whittier six months before the earthquake explains the urgency of Charles Woodworth's request for Smith's help in finding another anti-mosquito worker. Quayle's final assignment was to prepare an overview of the California anti-mosquito movement. During the fall, while Smith and Henry Clay Weeks were exchanging blows in *Entomological News*, Quayle completed his fifty-five-page report on the anti-mosquito movement in California.

The California Experiment Station published Quayle's *Mosquito Control* ten weeks after the earthquake. The report was the experiment station's first bulletin on mosquitoes. Like New Jersey, a desire for comfort and improving property values were the incentives for mosquito control in Burlingame and San Rafael. "This region," Quayle explained, "is essentially a country of homes, and many of the people of wealth. . . . They come here in order that they may live out of doors, and with hordes of hungry mosquitoes waiting to devour them . . . the desirability of inaugurating a warfare against the pests can at once be appreciated." A single season's anti-mosquito work "had a direct influence upon the real estate of the district, and many people were induced to locate in the territory largely on the basis that the mosquito nuisance was solved."[2] Like Hiram Byrd, Quayle was optimistic about the future of the mosquito crusade.

Woodworth searched three years for Quayle's replacement. In 1908 he took a chance and hired a thirty-two-year-old entomologist named William Herms. It was an improbable choice. Herms had published only three articles and lacked practical experience in mosquito control. Woodworth, however, believed that the young man's enthusiasm and evident passion for teaching compensated for his academic and practical shortcomings. He never regretted his decision.

Born in 1876 in Portsmouth, Ohio, William Brodbeck Herms overcame substantial hurdles to become one of the nation's leading medical entomologists. After graduating from high school in 1894 Herms worked for four years on local farms and as an accountant until he had saved enough money to attend nearby Baldwin-Wallace College in Berea, Ohio. During his freshman year Herms learned of Ross's discovery of the human-malaria-mosquito cycle. He later said, "It turned the tide of my life."[3]

At the turn of the twentieth century, malaria was a constant companion for those living in the Ohio River valley. Numerous folk traditions offered ways to prevent the disease. Herms retained a vivid memory of his father ordering him to be home before dark and being "frequently called down for eating too much watermelon and going out in the hot sun" lest he contract malaria.[4] Despite his scrupulous effort to follow his father's dictums, Herms suffered from a prolonged, seven-year struggle with the disease.[5]

Initially Herms planned to become a physician. His goal was to fight malaria as a medical missionary in China. By 1902, when he graduated from Baldwin-Wallace, his passion for entomology had eclipsed his interest in medicine. During the next six years, Herms alternated between periods of graduate study and spells of teaching in Ohio. He received fellowships from Western Reserve and Ohio State, where he earned his master's degree in 1906. In

1906–1907 he served as the acting head of the zoology department at Ohio Wesleyan College. His desire to go to China and fight malaria remained intact. The means of achieving this end had changed. While at Ohio Wesleyan, Herms had corresponded with John Smith. Smith offered to send him all of his mosquito publications. Gradually Herms came to believe that controlling mosquitoes offered a practical solution to the malaria problem.[6]

Herms spent 1907–1908 as an Austin Fellow studying zoology at Harvard. Herms, his wife, and their young son returned to Ohio at the end of the academic year. Two of his former professors, Herbert Osborn and E. W. Berger, recommended his appointment at Baldwin-Wallace College to teach accounting and biology. During the summer, however, Herms received a letter from Charles Woodworth offering him a position as an assistant professor of entomology at Berkeley at a salary of $1,500 per year. Herms's Harvard professors advised against accepting the appointment. They "shook their heads" and told him he "was taking an awful chance."[7] The University of California was "a dead end academically and the population of California was uncultured." Herms could do better for himself and his young wife. He should stay in Ohio or return to Cambridge and complete his doctorate. "Fortunately," William Reeves, who was Herms's student in the 1930s, observed, "Herms rationalized that California was much closer to China than Ohio and Massachusetts and he might still become a missionary selling malaria and mosquito control in California."[8]

Herms retained a vivid memory of his arrival in Berkeley. It was a cold, foggy August day. His first visit to campus added to his depression. "The small rather creaky wooden building known as the Entomology Building" was a disappointment. "I almost felt," Herms later observed, "that it was all a gloomy mistake."[9] There were no courses in the university's catalog on medical entomology. Woodworth and Quayle's work had focused on salt marsh mosquitoes. "I must confess," Herms wrote forty years later, "that my interest in salt-marsh mosquitoes was very mild—they were not vectors of malaria or of any other disease."[10]

Six months after his arrival, Woodworth invited Herms to join him on the experiment station's Agricultural and Horticultural Demonstration Train during the spring semester of 1909. The demonstration train was a joint effort of the University of California and the Southern Pacific Railroad to inform the public on agricultural topics. Herms's contribution to the entomology exhibits consisted of a "four small glass-covered boxes containing specimens of fleas, lice, flies, ticks and particularly specimens of anopheline mosquitoes, and a few charts."[11]

Herms's experiences on the agricultural demonstration train provided a basis for much of his future work. As the train made its way from southern to northern California, Herms came to appreciate California's diversity. The small farming communities in the Great Central Valley reminded him of things he had seen in the Ohio River valley. Stretching 430 miles from Redding in the north to the Tehachapi Mountains in the south, the hot, arid Central Valley did not physically resemble his childhood haunts. "Having been brought up in a malarial section of the country and still having the taste of quinine in my mouth," he observed, "my interest in malaria was intense." Many of the people he encountered carried the telltale signs of the wasting effect of malaria he had seen growing up in Ohio. Herms found clear evidence of malaria in the Sacramento Valley and in the foothills of the Sierra Nevada Mountains. Here Herms's brief remarks about medical entomology regularly elicited questions "concerning mosquitoes and malaria."[12] He returned to Berkeley convinced that malaria constituted a significant problem in parts of California. "A determination was awakened in me," Herms later declared, "to tackle the 'job' of control."[13]

Like Herms, malaria was a relatively recent arrival in California. Eighteenth- and early nineteen-century Spanish and American explorers and trappers make no mention it. Cases of intermittent fevers and chills were so infrequent in Southern California in the 1840s that a French visitor noted that upon hearing of a chill "that people traveled many miles to see [the sick man] shake."[14]

Fur trappers from the Hudson Bay Company probably carried malaria from Oregon into California in the early 1830s. There is wide agreement that a malaria epidemic ravaged the native population in the Sacramento Valley in 1833. "A terrible pestilence, an intermittent fever often prevalent in the region, is reported as having almost depopulated the whole valleys of the Sacramento and San Joaquin in the autumn of 1833," the nineteenth-century American historian George Bancroft wrote in his history of the West.[15] Demographer Sherburne F. Cook, who began pioneering studies of the impact of disease on aboriginal Americans in 1940, estimated that 20,000 Native Americans died during the epidemic.[16]

The discovery of gold in 1848 sparked a wave of immigrants to California. Some arrived in wagon trains. Others came by ship, crossing the Panamanian isthmus on foot and boarding another ship for the Pacific leg of the journey. Still others sailed from Europe and the East Coast around Cape Horn. The gold seekers introduced new sources of malaria into California. The crowding and poor sanitation in the squalid gold camps magnified the risk of an outbreak. An article in the *New York Times* in the late 1850s described malaria as a "growing evil in the mines." "The white man," the *Times* reporter declared, "cannot work for several months in summer, in the vicinity of these reservoirs."[17] Seventeen years later a correspondent for the *Chicago Daily Tribune* described certain counties in the San Joaquin and Sacramento valleys as places "where the quinine bottle is a constant companion."[18]

When Herms returned to Berkeley, Woodworth again suggested that he resume Henry Quayle's campaign against salt marsh mosquitoes. After the earthquake, the Improvement Club had not continued Quayle's initiative. Herms demurred. He wanted to work on malaria. Even his friends thought that he was making a mountain out of a molehill. The statistics supported this perspective. The overall incident rate for California was 4.9 deaths per 100,000 in 1908, as compared to 4.8 deaths for the entire nation.[19] This was insignificant when compared to the malaria rates for Florida, Mississippi, Alabama, and Georgia. Herms believed these numbers concealed the true magnitude of the problem. The challenge was proving it.[20]

On December 22, 1909, Herms received a letter from a young clergyman named Fred Morgan who lived in Penryn, California. "We, here in the Placer foothill region, want to fight the malaria mosquito, but we do not know how to proceed. . . . Can we not expect some aid in this?"[21] Residents of Placer County had long struggled against malaria. Thirty years before Morgan's letter, a reporter for *The Placer Press* in nearby Auburn lamented, "almost everyone living west of Gold Hill is either down with fever, or chills and fever, or more or less affected by the miasmatic poison generated and floating around in that locality. . . . What can be done to remedy the evil?"[22] A decade later a visitor to the gold mines reported seeing "many miners lying in rows at the bunk houses, stricken with malaria."[23] The mines had closed by 1880, casualties of the dwindling amount of bullion and the ravages of malaria.

Outbreaks of malaria were episodic during the gold rush era. This changed in the late 1880s. Farms and commercial fruit groves replaced gold mining as the county's principal source of income during this period. Irrigation made the country bloom. But the changes were not all beneficial. Most of the irrigation work was haphazard, creating ideal larval habitats for *Anopheles* mosquitoes. Poorly planned roads and railroad construction compounded the problem. "After long and sad experience," Herms and Harold Gray observed in 1940, "the writers are convinced that it is only by accident that either a railroad engineer or a highway engineer places a culvert properly."[24] In rural California, ineptitude and avarice had transformed a landscape "which previously had been healthy" into one that was "thoroughly infected with this curse [malaria]."[25]

J. Parker Whitney, a large landowner, sought to revive the county in 1890 by recruiting new settlers from abroad. He planned to breakup his two thousand acre estate into small farms and orchards. The brochures describing

California's pleasant climate and rich land attracted hundreds of English immigrants to Whitney's Placer County Citrus Colony. Whitney envisioned a prosperous, bustling community rising up in the foothills of the Sierras. He ordered the construction of tennis courts, a cricket playing field, and a granite clubhouse for the newly formed Citrus Colony Club. Whitney went so far as to organize "cross country paper chases . . . [that] fairly faithfully represented the fox hunts of Old England."[26]

Frank Kerslake, one of the English settlers, opened an agricultural college one mile west of Penryn. In 1891 and 1892 Kerslake "made frequent trips to England, sending out young men to take agricultural courses at this institution."[27] Nearly fifty students were enrolled in courses by 1893. The campus included a four-story classroom building, the head master's house, a stable, and other buildings.

The college students and English gentlemen farmers were no less susceptible to the intermittent fevers and chills that wasted their predecessors. Whitney and Kerslake watched in horror as the students and English farmers fled. The "English colony" with its foxhunts and cricket matches disappeared, the victim of the combined effects of the economic Panic of 1893 and malaria. In the late 1890s absentee property owners hired immigrant Chinese laborers to work on the Placer County Citrus Colony's abandoned farmland.

Fear that Chinese immigrants would supplant the rural white population in the Central Valley intensified as irrigation projects reshaped California. "Between 1900 and 1910," historian Linda Nash noted, "the length of main irrigation canals in the state went from 5,100 to 12,600 miles, and this did not account for all the minor canals and laterals that now laced the land."[28] In Placer County, Harry Butler, the manager and principal owner of the Penryn Citrus Company, made the connection between race, irrigation, and malaria explicit. "From the time the first orchards were planted (on a commercial scale about 1880) and irrigated from miners' ditches up to about 1910," Butler told Herms, "malaria was very bad indeed here. As a matter of fact we could not keep any white men at work. They had malaria and would leave as soon as able to walk."[29]

Racism played a pivotal role in launching anti-malaria work in California. Butler and other white settlers feared that Chinese laborers would replace the white population. At the same time, Butler blamed immigrants for the increase in malaria in the Central Valley. Harold Gray reported that Butler "and various old time residents of Loomis and Auburn . . . attributed the intense and pernicious malaria to the Chinese, Italian, and foreign laborers employed in the construction work."[30] Later Herms employed racist arguments to justify the need for a statewide action against malaria. "There are many localities in

California," Herms declared, "where malaria is virtually the only blight, and these include some of our most fruitful and beautiful portions of the state, made almost unhabitable [*sic*] for the white man, and now having largely been turned over to the Oriental, who apparently can get along, tolerably well at least, though infected with malaria."[31]

Much later, Stanley Bailey, one of Herms's students who went on to lead the entomology department at UC Davis, recast the racist argument in environmental terms. "It seems to be a characteristic of man—white man in particular—that wherever he goes he alters the environment. He clears the land, drains swamps, impounds water, and digs up the minerals thereby destroying wild plants and animals." White farmers, road builders, and railway men transformed an environment largely inhospitable to anopheline mosquitoes into an ideal habitat. "The changes and disruptions in the natural environment," Bailey concluded, "wrought by the white man resulted in marked fluctuations in the relative abundance of mosquitoes in mining and farming areas."[32]

Economics provided a second incentive for lowering the malaria rate. Malaria lowered productivity, drove up wages, and depressed land value. In 1918 Herms estimated that malaria cost the state nearly two million dollars annually, accounting for the loss of 168,000 workdays. "With malaria," Herms warned, "unless steps are taken for its control, its subtle and insidious attack upon the vitality, initiative, and enthusiasm of a people increases month by month and year by year until we find communities otherwise blessed with all the bounty of nature reduced to a collection of hovel and people by a race of anaemics."[33]

Fred Morgan was one of the white settlers who remained in Penryn. His position as a minister of a local church gave him a unique perspective on how malaria sapped the town's strength. The idea of asking for help in waging war on malaria grew out of a conversation between Morgan and Harry E. Butler in the fall of 1909. According to Butler, Morgan appeared at his office wanting to do something about malaria. "I've heard," Morgan declared, "a person can get malaria in no other way than from the bite of a mosquito. What do you know about it?" Butler had heard the same thing. If it was true, Butler declared, "why don't we do something about this?"[34]

Morgan's appeal arrived at an opportune moment. William Snow, the newly appointed Progressive secretary of the State Board of Health, issued a bulletin in December 1909 titled "Malaria the Minotaur of California." Evoking ancient Athens, which annually paid tribute to the Minotaur's bestial appetite, Snow contended that the politicians and absentee landlords had allowed the blight of malaria to ruin countless lives. "Throughout the length of the great Sacramento and San Joaquin valleys," Snow declared, "the Spirit-of-the-Swamp demands a tribute of those who would settle in its domain."[35]

Sadly, even those charged with protecting California's health had disregarded the growing malaria problem. The state health board had ignored malaria for nearly twenty years. Between 1892 and 1909, malaria received "scant attention" in the board of health's biennial reports.[36] "Malaria," Snow wrote in the board's twenty-first biennial report (1908–1910), "has taken its toll of human lives amounting to 192 for 1908 and 1909."[37] The Minotaur, Snow declared, can be slain "if the citizens of any malarious district will believe the facts of Science and act accordingly." Until they do so, a significant part of California would continue to resemble the "pen picture of human misery and inefficiency" that Charles Dickens depicted of a "malarious district" in *Martin Chuzzlewit*.[38]

Penryn was the first community to organize a campaign against the "Minotaur of California." Herms had immediately replied to Morgan's call for help. Noting John Smith's success in New Jersey, Herms advised Morgan and Butler that "a campaign such as you wish to undertake should begin as an educational movement."[39] He suggested that they meet on January 5 in Penryn and draw up plans for the campaign. Herms made weekly visits to Penryn over the next two months. Butler and Morgan organized public meetings in Penryn, Loomis, Newcastle, and Auburn where Herms gave speeches on malaria and mosquito control.

The Penryn mosquito extermination committee was formally organized on February 12, 1910. Herms outlined the proposed campaign at this meeting. He planned to concentrate the work in an area stretching from Newcastle in the west, through Penryn, and on to Loomis in the east. Harry Butler, the committee's permanent chairperson, supplied Herms with a room on his citrus farm to use as his laboratory and command post.

The campaign began on March 1. Herms discovered that there were large numbers of anopheline mosquitoes throughout Placer County. There were two principal sources for the anopheline problem in Penryn. The first grew from defects in the irrigation on local farms. The Southern Pacific Railway line that crossed the county had also created numerous additional drainage problems. Stagnant pools regularly formed in places where the railway grade blocked existing natural outlets. Solving Penryn's malaria problem meant that Herms had to persuade local farmers and the Southern Pacific Railway to join the mosquito crusade.[40]

Herms's resources were meager. The project was funded through donations. During the first year, the citizens of Penryn contributed $715 to the anti-mosquito campaign.[41] Many of the local people scoffed at the idea that something as inconsequential as a mosquito could cause malaria. Herms dismissed as ignorance the local belief that drinking stale beer caused malaria. It was more difficult to ignore the numerous yellow handbills that Herms found plastered around the county on the eve of a town hall meeting he had organized. The handbills declared "Don't go to hear that Professor from Berkeley. Every time you stub your toe on a gold nugget, it don't mean you will get malaria."[42]

Conservative chambers of commerce were formidable foes. Herms, Butler, and Morgan were "loudly denounced by certain boosters and certain newspaper editors" for creating "the wrong kind of advertising" for Penryn.[43] Merchants feared Herms's lectures would slow growth and bring Placer County into ill repute. Herms, a gifted teacher, confounded his critics. He argued that the town and county's future depended on the anti-malaria campaign. The true disgrace would be to continue to allow the "Minotaur of California" to lay waste to their community.

The long-term success of the Penryn project depended on Herms's ability to educate the people. He knew firsthand that without the community's support, the experiment would fail. While a student at Harvard, he had observed the limited success of one of the nation's earliest anti-malaria initiatives in nearby Brookline, Massachusetts. In 1901 "citizens who lived in one of the residential parts of the town" presented the local health board with a petition requesting help in ridding the community of mosquitoes.[44]

Complaints about the prevalence of mosquitoes in Brookline and Boston date from the early colonial period. It was not, however, until the beginning of the twentieth century that the residents in one of the town's well-to-do sections mounted a campaign for mosquito abatement. The petition noted that malaria had become a significant problem after Italian immigrant laborers were "employed upon the construction of the park system." The citizens committee demanded that "something . . . be done to safeguard the town from malaria."[45]

In 1902 Brookline's health board began anti-mosquito work without securing the community's support for the experiment. "We were to work miracles," J. Albert Nyhen, Brookline's health officer, reported. The people believe that "by simply waving of empty hands, the mosquitoes would disappear." Nyhen enlisted the support of Professor Theobald Smith at Harvard's medical school. Regrettably, city officials allocated "ridiculously small sums of money" to implement Smith and Nyhen's "mosquito suppression." "That was very wrong," Nyhen declared twenty years later, "but this notion has stayed with us even to the present time to the detriment of good field work."[46]

The amazing thing about the Penryn project was the miracle Herms accomplished with the "ridiculously small" sum of $715. His approach was innovative. "I recommend that at first a vigorous campaign be made in some restricted isolate area where *Anopheles* mosquitoes and malaria have been extremely bad," Herms explained to Butler and Morgan, "in order to demonstrate beyond question the practicability of the plan."[47] Earlier work like Alvah Doty's efforts in 1901 on Staten Island and in Brookline employed oil to reduce the entire mosquito population. These temporary measures were aimed indiscriminately at all species of mosquitoes. The numbers of anopheline mosquitoes, the vectors for malaria, fell as part of a generalized reduction of the overall mosquito population. The Staten Island and Brookline anti-malaria projects necessitated regular applications of oil to suppress mosquitoes.

The inspiration for Herms's Penryn campaign came from Malcolm Watson in distant British Malaya. Watson, a young English physician, arrived in the Federated Malay States in 1901 in the midst of a raging malaria epidemic. "I found the hospital at Klang," Watson later wrote, "full of malaria. It appeared to me that my duty consisted in doing more than remaining in hospital all day treating patients, since to this there could be no end if steps were not taken to prevent the infection of the population."[48] With no entomological training, Watson proceeded empirically "mapping out a programme of mosquito control by felling the jungle and draining the valleys and swamps around the town."[49] The malaria rate plummeted thereafter because *Anopheles umbrosus*, the malaria vector, did not oviposit in direct sunlight.

Watson's accomplishment was the recognition that "control must be selective, that it must take account of the breeding habits of the vectors and be directed specifically against them. Other species can be left alone."[50] By targeting the species of mosquitoes that was responsible for transmitting malaria, Watson discovered a practical means to eliminate disease. The Dutch entomologist Nicholas Swellengrebel called Watson's discovery "species sanitation." This was a strategy that could be applied in the mangroves in Klang or in the clogged ditches in Placer County.

Herms patterned his detailed plan for an eight-mile area surrounding Penryn on Watson's work. He identified the areas with the highest concentrations of anopheline mosquitoes. Most of his $715 budget went to ditching and draining these sites. His persistence, sincerity, and patience paid handsome dividends. Local farmers agreed to change their watering and irrigation practices. "Education was the first requirement," Herms discovered in Penryn. "California needed education along the lines of mosquito control. I began writing and publishing articles on mosquito control." Victory was certain if he could win the people's support.[51]

Convincing the Southern Pacific Railway to make expensive improvements posed a different challenge. Herms's experience on the Demonstration

Train gave him contacts in the company. He wrote and asked the railroad for their help. The company sent a roadmaster to meet with Herms. Herms took the man to Auburn and showed him how the railroad's grade created a drainage problem. Surely, Herms argued, a great corporation would do its part to free Penryn from the "Minotaur of California."

There were disappointments. Newcastle dropped out before the campaign's official beginning on March 1, 1910. Support in nearby Loomis waned in April and work stopped in May. Herms told Butler and Morgan that winning the war against malaria would take time. Butler, the campaign's permanent chair acknowledged this. "While the campaign was not enthusiastically supported in the district at first," he explained, "before the season had progressed far, much interest was manifested and at the close of the year sentiment in the Penryn district seems to be strongly in favor of continuing, on broader lines, such as the formation of a sanitary district."[52]

Herms's steady stream of articles, lectures, and demonstrations sparked statewide interest in the mosquito crusade. Herms received calls for help from other communities before the Penryn work was completed. Business leaders in Oroville in nearby Butte County asked for Herms's assistance in organizing an anti-mosquito league. In August the Women's Club in Bakersfield in distant Kern County appealed to him for guidance for their "anti-mosquito committee."[53]

Penryn provided a model for Herms's work in Oroville. Located on the Feather River, Oroville was a prosperous mining town with four thousand residents. Herms discovered that the malaria problem in Butte County was even greater than Placer County. The malaria death rate in Placer County in 1909 was 27.7 per 100,000. The death rate in Butte County was nearly three times greater with 64.3 deaths per 100,000.[54] Herms convinced the Oroville authorities to form an anti-mosquito league. As he had in Penryn, Herms agreed to serve as the city's unpaid deputy health officer. The City Council allocated six hundred dollars to his anti-mosquito work. Schoolchildren organized two "Tag Days" in which residents received an anti-malaria pin for a ten-cent contribution to Herms's "Benefit Mosquito Fund." William Reeves, who later served as dean of the UC Berkeley School of Public Health in the 1960s, contended that the Oroville ten-cent mosquito tag drive provided a model for Franklin Roosevelt's National Foundation for Infantile Paralysis Program. Reeves maintained that the Oroville mosquito tag drive "was the [nation's] first 'March of Dimes.'"[55]

The *Oroville Register* and the *Oroville Mercury* supported the anti-mosquito work. "The citizens of Oroville are to be congratulated," an editorial writer declared, "upon their wide-awake interest in matters of public health.

Outside interest is keen, and men of wide business reputation do not hesitate in expressing their praise with reference to this splendid movement. Oroville's example will be followed soon in many parts of the State."[56] The next year Thomas F. Means, manager of the Los Molinos Land Company, asked Herms to organize an anti-malaria unit in Tehama County.

The first season's results in Penryn, Oroville, and Bakersfield were impressive. Using school attendance records, Herms estimated that his "species sanitation" work had reduced the number of cases of malaria by nearly 50 percent.[57] The Kern County campaign had begun too late in the mosquito season to have a significant effect. The Bakersfield women's "anti-mosquito committee" raised one thousand dollars for the next season's campaign.

Herms delivered his report on Penryn, Oroville, and Bakersfield anti-mosquito work to William Snow and the state health board in November 1910. He advised that "communities contemplating anti-mosquito or anti-fly or anti-parasite crusades" begin their campaigns in the winter. The single greatest hurdle to "anti-mosquito organization in California" was the "unsatisfactory manner of raising funds by subscription to carry on this work." Herms recommended that California adopt some form of state legislation as did the "the legislature of New Jersey by an act approved on April 20, 1906" that would fund the anti-mosquito work. "New Jersey," Herms declared, "has found this a wise investment." California should follow New Jersey's lead in the battle against mosquitoes. "What this state [California] wants is the colonization of the great inland valleys and productive foothills of the Sierras," Herms maintained, "and the control of malaria is one of the great problems to be solved. But the problem is one that can be solved, as we are coming to see, and that at a reasonable cost."[58] Herms was convinced that in California and throughout the nation that "successful malaria control . . . is approximately synonymous with mosquito control."[59]

While Herms was hard at work in Penryn, John B. Smith was fighting to save the New Jersey crusade. Smith had high hopes for the anti-mosquito movement's future in 1906. The Duffield Amendment in 1904 had given local health boards the legal grounds to compel property owners eliminate "waters in which mosquito larvae breed." The 1906 salt marsh law addressed the problem of migratory mosquitoes that plagued two-thirds of the state. Smith had not anticipated, however, the growing reluctance of the appropriation committee to support the anti-mosquito work.

Initially, the passage of the 1906 salt marsh bill appeared to represent a tremendous advance. Editorials praising New Jersey's resolve in taking on the salt marsh problem appeared in the national press. Less than a week after Governor Stokes signed the salt marsh bill, a reporter for the *Chicago Daily*

Tribune declared that it was time for Chicago and Illinois to follow New Jersey's lead in the war on mosquitoes. "Chicago has many improvement associations," the correspondent declared. "So far as appears from any published report none of these of these associations has made or proposes to make a determined effort to exterminate the mosquito." Citing the success of Smith and Doty's work on Staten Island the reporter concluded, "the mosquito, like the cheap politician, is accepted by most people as a necessary evil. Yet the mosquito is not only more dangerous but easier to get rid of than the politician." All that was lacking was the political will to do what New Jersey had already done. "A little kerosene, a little drainage, a little care," the article concluded, "and the joy of living in August and September would be doubled."[60]

Less than a year after winning Edward Stokes's approval of the salt marsh bill, Smith found himself struggling to convince the appropriations committee and the governor to fund the needed work. "I think," Smith told Robert Engle, the owner of a Beach Haven resort hotel, "the Governor exaggerates the amount of censure that would be made if he approves the bill."[61] In 1907 Smith received only ten thousand dollars in the regular appropriation bill for work at the experiment station. In the weeks before the legislative session ended, he succeeded in winning an additional ten thousand dollars in a supplemental appropriation bill. Doty and Smith's limited ditching plan for Staten Island had cost $17,000. Addressing the mosquito problem in the Garden States 296,000 acres of salt marsh required a much greater sum.[62] The $350,000 needed to remedy the salt marsh problem remained illusive.[63]

Ironically, Smith's political problems in Trenton materialized when his anti-mosquito work was receiving accolades in the nation's newspapers. "People in Jersey City are sitting on their verandas this Summer," a reporter for the *New York Times* declared, "without inclosing screens—an unusual luxury. Golf is played in Richmond without gloves and veils. People can wear Oxford shoes minus spats. And the end is not yet."[64] Alvah Doty declared that *Ae. sollicitans*, "the striped legger" had "practically disappeared from Staten Island." "The experiments," Doty declared, "have succeeded beyond . . . expectations." If the work was continued, Doty believed that it might be possible to "accomplish the extermination of the whole miserable family eventually."[65]

The Staten Island work was completed in August 1907. In less than a year Doty and Smith's mosquito brigade had dug more than four hundred miles of canals and ditches in a twenty-square-mile area. The final segment of the lattice work of two-feet-deep and twenty-inch-wide ditches were dug in the salt marsh along the Fresh Kill on the west side of the island. The *New York Times* congratulated Doty and Smith noting that "very little oil" was being used. Soon Staten Island and New Jersey, the *Times* reporter declared, would be free of the mosquito pest. "The Jersey mosquito," the *Times* reporter declared, "is on the way to extinction as surely as the bison and the American Indian."[66]

The newspaper's comparison of the mosquito crusade to the "extinction" of the "bison and American Indian" is revealing. It would haunt the anti-mosquito movement. In their effort to win converts, Smith and Doty had created the unfortunate popular opinion that the only good mosquito was a dead mosquito. Five years earlier Smith had declared his objective was not extermination. His goal was reducing the number of mosquitoes to a level that they would no longer be a nuisance.

Smith and Doty discounted the press's ballyhoo. Certainly, the Staten Island work was a positive development. It demonstrated the soundness of Smith's *Mosquitoes of New Jersey*. The mosquito crusade was attracting new recruits. In December 1907 Smith received a copy of Evelyn Mitchell's *Mosquito Life*. Mitchell, a Cornell University–trained entomological illustrator, was the first woman to write a book on mosquitoes. Three years earlier Dr. James Dupree, surgeon general of Louisiana, and Harcourt Morgan, a professor of zoology at Louisiana State who later became the president of the University of Tennessee and one of the directors of the Tennessee Valley Authority, had written to John Comstock at Cornell asking for help in locating an illustrator. Comstock had recommended Mitchell, who had spent six months working with Dupree. Mitchell developed a series of identification keys for adults, larvae, and mosquito eggs during her stay in Louisiana. At the end of six months Mitchell returned to Washington where she completed her master's degree at George Washington University. After graduation she found work as an artist entomologist in the Division of Medicine at the National Museum.

In 1905 Mitchell decided to expand her master's thesis on the Louisiana keys. She asked for Dupree's assistance. He volunteered to supply her with his notes "knowing that he need not worry about receiving proper credit." His death in April 1906 came before Mitchell had completed her work. Fortunately, Col. Thomas Boyd, president of Louisiana State University, and Morgan, who had left LSU to become Tennessee's state entomologist, agreed to send Mitchell all of Dupree's notes.[67]

Mitchell sent Smith a copy of *Mosquito Life* in November 1906. "I took up your book on *Mosquito Life* last night intending to glance it over," Smith wrote, "and found it so interesting that I stuck to it until I had read it all except the tables. I think that you have succeeded excellently in your purpose of presenting the whole subject in concise form and in such shape that any one interested in the subject at all can get a comprehensive view of it." Like Herms, Smith believed that the success of the anti-mosquito movement depended on education. Smith valued Mitchell's *Mosquito Life* because the book had "given all that is necessary of structural details" in a manner that did not "tire or repel the general reader." This was an important accomplishment.[68]

Smith received Mitchell's book at the beginning of a series of setbacks in the mosquito crusade. During the next four years, budget cuts forced Smith to limit his anti-mosquito initiatives. New Jersey's newly elected governor in 1908, Franklin Fort, and the leaders of the legislature opposed increasing appropriations for mosquito work. Worse news lay ahead.

"The Legislature at its session in 1909 made no provision in the regular appropriation bill to continue the work of mosquito extermination, under Chapter 134 of the Laws of 1906," Smith informed Edward Voorhees, director of the experiment station. "It has been necessary, therefore, since November 1, 1909, to discontinue active work."[69] The budget cuts' effects were immediate. Smith had to dismiss his cadre of mosquito workers. He tried to find jobs for John Grossbeck and Hermann Brehme. John Grossbeck left New Jersey and found work in Illinois. Brehme continued work on an unpaid basis. "I should hate to lose him," Smith observed, "but I am at the mercy of the legislature, and can never tell from year to year what I am going to have. I never like to stand in the way of a man bettering himself, and if you can make Mr. Brehme a proposition . . . I will be pleased."[70] The cuts had a significant effect on Smith's ability to help local health boards. Since 1901, local health boards had depended on Smith to provide scientific assistance. "I am sorry to say," Smith wrote to John Dobbins, secretary of the Newark Board of Health, "it is utterly impossible for me to do anything in this line at the present time."[71]

The detection of brown-tailed and gypsy moths in New Jersey presented an additional problem. Foresters first noted the arrival of these invasive species in New England two years earlier. Unopposed, the gypsy moth, *Lymantria dispar*, can defoliate entire forests. Smith spent much of 1910 organizing efforts to eradicate these insects. One positive aspect of the gypsy moth invasion was that the legislature and governor agreed to renew Smith's appropriations. This allowed him to rehire Hermann Brehme.

The New Jersey anti-mosquito movement reached a watershed in the summer of 1911. The leaders of the anti-mosquito movement in northern New Jersey formed a new association in 1910, the North Jersey Mosquito Extermination League.[72] The Mosquito Extermination League's monthly meetings initially served as a forum for local health boards to meet and discuss shared problems. The League's members found themselves facing a tremendous challenge in 1911. May and early June had been unseasonably dry. On June 5 and 6, heavy rains fell all across New Jersey. Simultaneously, a period of high tides flooded many areas that had remained dry for the two previous years. By the end of June clouds of mosquitoes were descending on Jersey City, Newark, Perth Amboy, Elizabeth, and numerous other cities and towns.

The 1911 infestation lent support to the idea that the anti-mosquito movement was ineffective. Critics claimed that the local and state campaigns to eliminate mosquitoes had little to show for their costly expenditures. Smith was alarmed when he learned that there was growing opposition for mosquito control work in Trenton. For several years William Bradley, the powerful chair of the Senate Appropriations Committee, maintained that there were more mosquitoes where the work was done than there were before.[73] If Bradley's opinion became widespread, the future of the mosquito crusade would be thrown into doubt.

Smith received a torrent of letters asking for an explanation. "It is one of those combinations that for the present," he replied to a city official in Perth Amboy, "we are helpless." The local health boards and members of the North Jersey Mosquito Extermination League had done what they could to prevent the infestation. "Let me again assure you," Smith continued, "that the conditions are not due to any neglect of your local authorities."[74]

The June 1911 mosquito invasion exposed two flaws in New Jersey's laws governing anti-mosquito work. First, there was no way to compel the cooperation of neighboring communities undertaking mosquito control. "It is unfortunate," Smith explained to South Orange's health officer, A. C. Benedict, "that the authorities will do nothing, and will maintain conditions that make it difficult for neighboring communities to secure relief."[75] South Orange, Newark, Jersey City, Elizabeth, and the twenty-seven other members of the Mosquito Extermination League were acting responsibly. Other towns had chosen to ignore the mosquito blight. "The whole truth of the matter," Smith believed, "is that most localities have not taken the mosquito problem in hand seriously enough."[76] There was general agreement that the future of the anti-mosquito movement rested on finding a way to encourage local communities to cooperate.

New Jersey's salt marshes posed an even greater problem. It was impossible for communities like South Orange or Jersey City to control mosquitoes outside their jurisdiction. Smith believed it was imperative that the state take on governance of the salt marshes.

The state had allocated nearly $83,500 for permanent control work along the north New Jersey coast since 1902. There was, however, no mechanism to protect the work. "It is necessary each year for the 'unfortunate mosquito man' to go to the Legislature and beg for money to continue his work," Smith explained to Dr. B. Van Doren Hedges in 1911. "One of the difficulties that he meets is that there is each year a different appropriation committee, and very frequently a set of men entirely new to the legislative mill and uninformed as to the work being done."[77] In 1910 the "legislative mill" had eliminated funding for the salt marsh work. This was the "weak spot" in the current legislation. "At the present time," Smith explained to Robert Engle, "we have no method of protecting the work done by the State, and this is one of the points that will have to be taken up in the next session of the legislature. Just what the outcome is going to be I am not sure. A committee has been appointed by the Municipalities in Union, Essex, and Hudson Counties, and I will keep you advised as to what their plans will be."[78]

A debate raged for six months within the anti-mosquito movement over the proper course of action. "The necessity of legislation," Smith wrote in a letter to Susanne Strong, a New Jersey Progressive, "is fully realized; but just exactly what legislation is needed and upon what communities the burden shall fall is a matter of discussion."[79] Strong responded "our Women's Clubs are a power if they can be aroused. . . . They have been interested in the question [of mosquito control] but they have never had a definite, concrete object before them such as they would have if you sent in a draft bill or some such thing which would work towards the complete extermination of that horrid pest."[80]

Smith collapsed before he could prepare a draft bill for the needed legislation. Years of overwork and the strain of his ongoing battles in Trenton proved too much for him. "My appearance at Mount Holly was my last public bit of work up to the present time," Smith confided to his friend Turner Brakeley in an October letter. "I kept up on that day on will and whiskey, and when I got back home, gave up. I went to bed after Labor Day and have just graduated from that institution, although I have not yet been down to the office."[81] Smith suffered from Bright's disease, an incurable kidney disorder. His doctor ordered him "to cut off everything that can be cut off, and give [gave] me very little hope that I will ever be able to do anything like the work I have been doing in the past."[82]

Smith's illness came at a critical moment for the mosquito movement. Smith had led the New Jersey mosquito crusade since its inception. He won funding in 1902 for his study, *The Mosquitoes of New Jersey*, and his experiments with controlling salt marsh mosquitoes. In 1904 he secured the approval of the Duffield Amendment, which gave local health boards the power to act against mosquitoes. Finally, in 1906 he persuaded the legislature and the governor to commit to a multiyear campaign to eliminate salt marsh mosquitoes. Smith's intelligence and political savvy were sorely missed in the next six months.

Three factions developed within the North Jersey Mosquito Extermination League. John Dobbins, secretary of the Newark Health Board and the league's president, argued for an expansion of the power of local health boards.

William Delaney, secretary of the Jersey City Health Board and chair of the league's legislative committee, contended that counties should have the right to form three-person mosquito commissions. Edward Ayers, a prominent physician, represented the third group that advocated the creation of a statewide mosquito commission to govern all anti-mosquito work.

Smith did not think the proposals represented an improvement. "The trouble is that you gentlemen do not seem to recognize exactly what is really needed," Smith declared in a letter dictated to Augusta Meske from his New Brunswick bed, "and what modifications and additions to the existing legislation [are] required. Such propositions as have been made up to the present are good as means to an end; but they are methods that are suggested and not organizations that are created to carry out policies."[83]

The debate within the North Jersey Mosquito Extermination League took an unexpected turn in October 1911. The *Washington Post* reported that the New England Society of New Jersey, an association of prominent individuals living in New Jersey who were born in New England, "started what looks to us like a subtle and malicious assault upon Gov. Woodrow Wilson with a view to breaking down his presidential boom. The society cloaks its attack under the popular cry, 'Death to the mosquito!' which every Jerseyman subscribes to." At its autumn meeting, the society had passed a resolution criticizing Wilson and condemning "the ridiculous position occupied by the State in that it does nothing as a Commonwealth to get rid of its worst enemy."[84]

During the society's meeting, speakers accused Wilson's "inaction" in New Jersey's campaign to eliminate mosquitoes. This criticism carried an implicit comparison to President Taft's reputation as a mosquito warrior. Taft's supporters pointed to his support of anti-mosquito work in the Philippines and Panama. The *Post*'s article ended with an editorial comment defending Wilson. "Is it fair, is it right, is it decent politics to make such demands on Gov. Wilson which are palpably out of his line of activity and beyond his power," the *Post's* reporter concluded. "Let the enemy take any other shape than a mosquito, and Woodrow Wilson would not flinch. But it is asking too much of him to demand that he shall rid New Jersey of mosquitoes."[85]

Smith was not amused by the *Washington Post*'s story. He wrote a scorching letter to John Dobbins, Essex County's health officer, when he learned that Dobbins and Spencer Miller had supported the New England Society's resolution. Dobbins claimed that he and Miller had not favored the wording of the resolution. They believed it would "serve to bring the matter to the attention of the Governor with the hope that the Governor would become interested and take some action that would be beneficial to the cause."[86] Smith disagreed. "I have seen the clipping from the 'Sunday Call' . . . and under the circumstances it seems to me that the less that I have to say or write to you and your various bodies the better it will be." The episode left a "bad flavor." "There is," Smith concluded, "plenty of criticism and plenty of loose assertion; but loose assertion will cut very little figure this year. It strikes me that what is being done now is allowing things to remain as they exist and trying to go off and wade through the mud in some different direction instead of making use of the knowledge we already have."[87] Smith's declining health, however, prevented him from leading the contending factions within the Mosquito Extermination League out of "the mud."

The debate between the Dobbins and William Delaney, Hudson County's health officer, intensified in December. Hermann Brehme kept Smith apprised to the heated exchanges between the Essex and Hudson factions that took place in the meetings of the league's legislative committee. "To tell the truth," Brehme confided to Smith, "they do not know what they want." On the one hand, Delaney argued that mosquito control should be organized around three-person county commissions. The commissions would submit their budgets to the freeholders within each county who would provide funds for the mosquito work. Dobbins, on the other hand, wanted to protect the Newark Health Board's near eighteen thousand dollar outlay for mosquito abatement. "I don't know how to take Delaney and Dobbins," Brehme declared. "In the first place Delaney is drawing up the bill and at the same time he does not want a commission in his own county for fear he will lose his position."[88]

Delaney and Dobbins had "axes to grind."[89] "I could not under any circumstances," Smith told Dobbins, "approve of the adoption of any set of resolutions like those sent to me." Neither man had addressed the salt marsh problem. There was no provision for additional funding in either proposal. At best, the proposals offered an organizational change. At worst, they posed a threat to the future of the anti-mosquito movement. At least this was Hermann Brehme's interpretation of the debate. As Smith's illness worsened, he increasingly relied on Brehme's reports on the ongoing debate. "It is a safe bet," Brehme wrote to Smith after witnessing an angry exchange between Dobbins and Delaney, "that if this bill is passed, then, mosquito work will be dead, in every county with [probably] Essex and Hudson excepted, as none of the southern counties will continue the work, if the money has to come from county taxes."[90] Smith agreed and warned Dobbins and Delaney. "If I cannot find a way out in the course of the next week or ten days," Smith declared, "I shall go to Trenton . . . and end up this whole business. At all events I can see the Governor and I can tell him how this matter stands so as to keep his hand off completely."[91]

The commission proposal was doomed if Smith carried through on his threat to meet with Gov. Woodrow Wilson. Delaney, Spencer Miller, and

Edward Loomis knew that Smith would wait until after Christmas before scheduling an appointment with the governor. They used the time to reformulate their proposal. Miller and Loomis asked Edward Duffield to review the draft bill. Duffield's political fortunes had improved in the ten years since he first ran for a seat in the Assembly. He had recently stepped down as New Jersey's attorney general. He remained a staunch supporter of the anti-mosquito movement.

Duffield proposed a significant revision in Delaney's draft. He recommended that the new proposal include language specifying that the director of the state experiment station was responsible for approving the abatement plans and budget proposals of the county commissions. In practical terms, this meant that the state entomologist would have final authority over the various counties' extermination programs.

Loomis sent Smith a copy of Duffield's revised version of the bill in early January. "I read the bill," Smith replied two days later, "as amended by Mr. Duffield with a great deal of interest. It is much narrower and much more specific than the previous legislation that has been submitted and may cover all the requirements."[92] Smith's response encouraged Miller, Delaney, and Loomis to go to New Brunswick and make a personal appeal for Smith's support of the commission plan.

"It is well known," Miller recalled years after the visit, "that Dr. J. B. Smith . . . was utterly opposed to the commission plan."[93] Smith remained doubtful about the commission plan. Duffield's revisions, however, had dulled his opposition to it. In mid-January, Smith sent a lukewarm endorsement of Delaney's proposal to the Mosquito Extermination League. "I regret sincerely," Smith declared, "my continued inability to meet with you. . . . I keep close to you of course, and in touch with all you do, and wish to express in this connection, my belief that the County organization for the mosquito fight as outlined in the bill recently submitted to me by Mr. Loomis, is a long step in the right direction. Whether it will prove the ultimate solution of the matter in the State it is perhaps too early to say."[94]

The legislative committee met the following evening. Hermann Brehme sent Smith a detailed description of the proceedings. "The meeting was a hot one," Brehme wrote. "Dobbins and Delaney pitched into one another vigorously, Delaney fighting for the bill and Dobbins against it. Dobbins threw a bomb in the fire when he stated that the Newark Board of Health are against the bill, to make matters worse he stated that the Essex County Board of Freeholders are against the bill as well." In the midst of the debate "the meeting was interrupted by the Janitor dropping dead." After the man's body was removed the committee voted to support Delaney's proposal.[95]

Figure 3.1. William Morton Jackson Rice, *John Bernard Smith* (1858–1912), 1913. Oil on canvas (76.2 x 63.5 cm). Reproduction courtesy of Jane Voorhees Zimmerli Art Museum, Rutgers, The State University of New Jersey, gift of Mrs. John B. Smith 0123.

Smith remained ambivalent about the proposal. "Obviously," he wrote to A. C. Benedict, "you had a lively and deadly time—all in one evening . . . I told Mr. Loomis as I have told everybody with whom I conferred that I am altogether dissatisfied with the present condition of affairs and with the lack of power that exists in doing work on a sufficiently extensive scale." That said, Smith declared that the "Duffield bill came as near to a solution of the problem as anything yet presented up to the present time; but I am not sure that it fills the bill even yet."[96] Smith was blunter in his comments to his friends. "I do not believe there is anything in either [Assembly and Senate versions of the commission proposal] that I want to fight," Smith told Franklin Dye, the secretary of the New Jersey Agriculture Board, "and, on the other hand, I do not believe that there is anything in either of them that is going to be practically useful; but I must know what is being done in order to protect the work that has been already completed."[97]

While the debate about the commission proposal raged in Trenton, Smith focused on his work at the experiment station. The long-planned new entomology building was nearly complete. Smith drew up detailed plans for the laboratories and teaching auditoriums. An unhappy beekeeper alleged that

Smith had unfairly reviewed his application for an inspector's position. Smith prepared a meticulous report rebutting the charges for the state civil service commission. He could not keep up the pace. "I have been burning the candle at both ends," he told his close friend Turner Brakeley, "and of course have to pay the penalty. In a way I do not mind it; but it is rather tedious waiting for the candle to sniff out."[98]

Smith died on March 9, 1912. Twelve days later Woodrow Wilson signed the bill authorizing the creation of mosquito commissions in New Jersey. Five days before Smith's death, Louis Richards, the health officer in Elizabeth, had written to Smith asking for his help. "Unfortunately," Smith replied, as matters stand at present time with me I can give you no help whatever. This is not a matter of feeling or desire on my part; but dead up and down misfortune. If I get out of this, I may be able to take hold again and help. If matters, as I anticipate, turn the other way it simply means that somebody else has got to take charge of the whole subject, whose policy may be a totally different one from my own."[99]

Five months after Smith's death a young entomologist named Thomas Headlee left Kansas State University and took charge of Smith's work in New Brunswick. Headlee guided the development of mosquito control in New Jersey for the next three decades. Three thousand miles to the west Herms had formed a cadre of mosquito warriors that would defeat the "Minotaur of California" and organize a statewide anti-mosquito movement. John Smith, "the unfortunate mosquito man," was dead, but his dream endured.

Chapter 4 Public Health, Race, and Mosquitoes

CHILDREN GROWING UP IN the rural South in the early decades of the twentieth century learned to pay close attention to nature. In the Mississippi Delta lands, frogs proffered medicinal advice. Folk traditions taught that when evening fell, little frogs from the Arkansas swamplands to the Virginia Tidewater Basin piped "quinine, quinine, quinine" to which "the big bullfrog with his deep voice replied, double the dose, double the dose."[1] Until the beginning of the twentieth century, except for the frogs' croaking advice, the mill workers, sharecroppers, and poor farmers from the Carolinas westward to the foothills of the Sierra Nevada Mountains in California received little assistance in their struggle against mosquitoes and mosquito-borne diseases.

The decade between 1912 and 1922 brought new energy to the mosquito crusade. During this period the annual meetings of the New Jersey Mosquito Extermination Association (NJMEA) provided a rallying point for the anti-mosquito movement. In the northeast, Thomas Headlee consolidated Smith's gains and led the effort to extend campaigns against pest and nuisance mosquitoes to New England and the Middle Atlantic states. In distant California William Herms fought a dogged battle to secure a law authorizing the formation of mosquito abatement districts. Simultaneously, the creation of the U.S. Public Health Service (USPHS), the Rockefeller Foundation's commitment to improving rural sanitation, and most importantly America's entry into World War I marked the beginning of a forty-year struggle to vanquish mosquito-borne diseases from the South.

Seven months elapsed before Jacob Lipman, director of the agricultural experiment station at Rutgers, recruited a successor for John Smith. Lipman offered the position to a thirty-five-year-old professor named Thomas Headlee.

Born in Indiana in 1877, Headlee earned both his undergraduate and master's degrees from Indiana University. After completing his doctorate in entomology at Cornell, Headlee served in 1906 and 1907 as the assistant state entomologist at the New Hampshire Agricultural Experiment Station. Headlee accepted a position as head of the entomology department at Kansas State Agricultural Experiment Station in 1907.[2]

Headlee assumed his duties as a professor and New Jersey state entomologist in October 1912. He had hardly unpacked when he found himself in the middle of a political fight over the future of the mosquito commissions. Headlee quickly learned that mosquito warriors alternated between summer battles with six-legged pests and the winter fights with two-legged, political foes around the state.

As Smith had predicted, Essex, Union, and Hudson counties were the first to form mosquito extermination commissions. Chapter 104 of the Laws of 1912 stipulated that the county supreme courts would choose the members of the commissions. The bill's authors hoped that this method of appointment would insulate the commissions from political pressure while encouraging the selection of qualified individuals. Each commission consisted of six members, one-half of whom were to be drawn from the local health board. The director of the state agricultural experiment station, or his designee, was named an ex officio member of each of the commissions. Chapter 104 also gave the experiment station's director or his appointee the power to approve or reject the commissions' budgets and operational plans. The law mandated that freeholders fund the budgets of mosquito abatement commissions that were approved by the experiment station's director.

Badge-carrying mosquito inspectors fanned out across Essex and Union counties in June 1912. The Newark press applauded the commission's professionalism. "In no previous summer season," a reporter at the *Newark Evening Star* declared, "were Newark and its suburbs so free from the pest as during the present season. . . . The money appropriated for this blessed work is the best public investment ever made in this county."[3]

During his first year in New Jersey, Headlee traveled more than fifteen thousand miles, making hundreds of presentations rallying support for the formation of mosquito extermination commissions. Headlee's talent for organization, innovative scientific techniques, and quick mastery of the intricacies of the New Jersey political scene led to the formation of eleven commissions by 1915.[4]

Headlee's first challenge was deflecting efforts to repeal the 1912 commission law. Opposition to the 1912 law came from three different sources. In Essex and Union counties a handful of property owners objected to the mosquito inspectors' entry onto their property. The success of the Essex and Union commissions in reducing pest mosquitoes blunted these objections. Municipal and county health boards represented a second potential source of opposition. Health boards that had initiated their own anti-mosquito campaigns were reluctant to lose control of projects that they had financed.

The freeholders (county supervisors) proved to be Chapter 104's foremost critics. They resented the provision granting the independent, unelected commissions the power to compel the freeholders to allocate funds for mosquito control. "The mosquito extermination act," the editors of the *Newark Evening Star* observed in August 1912, "has been amply justified by results and its repeal by the legislature at the demand of some parsimonious county that is willing to suffer the pest rather than pay the small price of getting rid of it is not possible."[5]

In 1913 the law's opponents sought to nullify Chapter 104 through "an innocuous amendment." Chapter 104 explicitly stated that counties "must" fund anti-mosquito work if the state entomologist approved extermination commission's budget. The proposed amendment substituted the word "may" for "must," removing the freeholders' obligation to fund anti-mosquito work. Headlee's political savvy and painstaking attention to detail paid handsome dividends.[6] Evidence of this came during the 1913 legislative debate about the commission law. Neither the Essex nor the Union freeholders supported the amendment, and the measure was defeated.

The repeal effort underscored the importance of vigilance and organization. In November 1913 Headlee organized a meeting in Camden to discuss the mosquito crusade's future. He proposed that the old North Jersey Mosquito Extermination League serve as the nucleus of a statewide anti-mosquito organization: the New Jersey Mosquito Extermination Association (NJMEA). The participants agreed and elected Dr. Ralph Hunt, an Essex County mosquito commissioner, as NJMEA's president.

The first convention of the NJMEA convened three months later on February 20, 1914, in Atlantic City at the Hotel Traymore. Spencer Miller was the first person to address the assembly. "When I read," Miller told the approximately forty attendees, "the names of men serving on the county commissions of the State, printed on the back of our program, I felt a glow of exhilaration. People of this sort are putting New Jersey at the head of the States of the Union in practical mosquito work." The mosquito warriors, Miller declared, must not lose sight of the "interdependence of the mosquito fighting units."[7]

William Delaney, whose efforts to win passage of Chapter 104 earned his appointment as superintendent for the Hudson Mosquito Commission, spoke next. Vigilance, Delaney cautioned, was essential. The recent effort to repeal the commission law demonstrated the opposition's strength. The enemies of mosquito crusade, Delaney maintained, fall into two groups: those who do not believe and those who do not want to believe. The latter were the most troubling. No matter how much the mosquito commissions accomplish, these individuals will remain the commissions' implacable foes. The anti-mosquito movement's only sure defense was an informed public.[8]

Headlee used his presentation to the association to review the benefits of mosquito control. There was general agreement that the reduction in the number of salt marsh mosquitoes improved the quality of life. What was not generally recognized was the economic value of the reduction of pest and nuisance mosquitoes. Headlee asked Hermann Brehme to research what effect anti-mosquito work had on improving the value of real estate. Though incomplete, Brehme's preliminary findings suggested that mosquito control brought a sizeable increase in the value of land. Assessed property values in Jersey City and Rumson had increased by $26 million since beginning anti-mosquito work.[9] Headlee estimated that a statewide adoption of mosquito control would lead to as much as a one billion dollar increase in the taxable rates. He predicted that the increase in real estate taxes would pay for the anti-mosquito work.[10]

The public's support for the mosquito crusade grew after the first year's work. In 1914 Headlee conducted a telephone survey in the counties with mosquito commissions. Each of the selected individuals was asked if he had benefited from the anti-mosquito work; 80 percent of the respondents declared that they could attest to the work's effectiveness and 95 percent "favored the continuance of the work."[11]

Real estate developers were destined to become the anti-mosquito movement's staunchest allies. Twelve years after the NJMEA's founding the *New York Times* reported that "real estate men in New Jersey" had a "keen interest" in mosquito control. "All of them," the *Times* reporter observed in its real estate section, recognize "that their holdings would be greatly increased if the mosquito menace was kept well within control."[12]

Ironically, the real estate lobby in California proved to be the most formidable adversary to the anti-mosquito campaign. In 1911 William Herms had hoped to build a statewide anti-malaria campaign after his success in Penryn and Oroville. Money was the principal obstacle to expanding the mosquito work. In the first year of the Penryn work, Harry Butler, the owner of the Penryn Citrus Company, donated most of the funds for the project. At the beginning of the second year, Butler "demanded that this year [1911] everyone pay his proper share."[13] Herms agreed. "In any malaria crusade," he wrote, "all the inhabitants of a given district are equally benefited; it is therefore unreasonable that the entire cost of a campaign should be borne by a few individuals."[14] John Smith's success in 1906 in winning state support for salt marsh work provided a model for Herms's anti-malaria legislation.

In 1911 Herms persuaded John Guill, a newly elected Democratic assemblyman from Butte County, to prepare a mosquito abatement bill for the upcoming legislative session. Guill, who had witnessed the debilitating effect of malaria in Oroville and Chico, introduced legislation that gave county supervisors the power to undertake "all work necessary for the extermination of *Anopheles* mosquitoes, describing the boundaries of the district to be benefited and assessed for the benefits." County supervisors were required to provide certification from the state board of health that there were anopheline breeding places within the county. If this condition was met, the supervisors could appoint three mosquito commissioners who would be charged with making a "thorough sanitary survey of the district" and preparing an anti-malaria campaign.[15]

A majority of California's assembly members and senators supported Herms and Guill's proposal. The bill, however, had strong opponents. Working quietly behind the scenes in Sacramento, the California real estate lobby persuaded Gov. Hiram Johnson to veto the bill. Developers feared that the creation of anti-malaria abatement districts would dampen real estate sales. The lobby's opposition "was paradoxical," Earl Mortenson, a trustee of the Contra Costa County Mosquito and Vector Control District, declared, "because real estate interests were pushing hard to develop irrigation districts in order to enhance the sale of farm land, and attract settlement in northern California, referred in the news media as 'Superior California.'" The paradox was that the "irrigation projects" that raised the land's selling price often created ideal habitats for anopheline mosquitoes. Companies would sell land in Superior California only to discover that "as a result of the malaria threat," as Mortenson noted, "settlers wanted their money refunded on their newly purchased farm land."[16]

Herms's work in Placer County documented this. Malaria had become endemic in the district only after the introduction of irrigation in the 1890s. "Permanent control of malaria in the vicinity of Penryn," Herms wrote in his report on the project, "depends on the control of waste irrigation waters by means of efficient drainage, the provision of overflows and waterways from main ditches and laterals, and the prevention of leaks and seepage from the ditches."[17] Herms contended that the real estate lobby's opposition to the Guill bill was misguided. Certainly, they were right to fear malaria. "Real estate values depreciate, "Herms observed, when a district has a reputation for being malarial. "A certain property in Penryn," Herms explained, "was several years ago listed for sale in San Francisco, but the agents frankly told the owner that it would be impossible to sell on account of the reputation of the region for malaria."[18] Ignoring the problem, however, only made it worse. Land developers were unconvinced.

Herms was philosophical about the defeat. "The time when real estate men denounce anti-malaria efforts out of fear of exposure is rapidly drawing to an end," he wrote.[19] There were hopeful signs. Six months after the governor's veto, Thomas Means, manager of the Los Molinos Land Company in Tehama County, asked Herms to prepare an anti-malaria plan for the company's irrigated lands. Because it was the beginning of the fall semester, Herms could not leave Berkeley. Instead he asked a young civil engineer named Harold Gray to go to Los Molinos.

Gray first came to Herms's attention a year earlier in September 1910. At the height of the malaria season, Herms's field inspector in Penryn quit. Gray, who had recently resigned as a draftsman for the Sacramento Valley Irrigation Company, needed a job. He possessed superb qualifications. He had earned his civil engineering degree from Berkeley in 1907 and worked as an engineering assistant for the Peoples Water Company in Oakland and as a surveyor on the Los Angeles aqueduct project.

Gray proved a tireless and resourceful mosquito warrior in Penryn. Herms wrote four years later in his preface to his first book on malaria, "Much credit is due to Mr. Harold F. Gray, who, through his excellent equipment and efficiency, has given indispensable assistance to the preparation of this work."[20] Gray's scientific and technical work was flawless. More important, Herms discovered Gray was "an educator and a good 'mixer.'" These qualities were of particular importance. The public needed to be convinced that mosquito control was possible. "Sanitary inspection," Herms later wrote, "is not so much the exercise of police power, which is indeed of very minor importance, but it is the process whereby a citizen is made to see defects and is shown the way to correct the same."[21]

Herms spent the summer of 1911 in Europe. In his official capacity as "Officer in Charge of Malaria Investigations" for the California Board of Health, he visited the malaria-ridden Roman Campagna and schools of tropical medicine in England, and he was California's representative at the International Hygiene Exhibit in Dresden. While abroad, Herms placed Gray in charge of organizing the second season's work in Oroville. Gray remained in Oroville for eight months. He devoted much of his energy to "educational publicity, and an earnest attempt to bring citizens to a realization of the importance of better health conditions." He told Herms "from the standpoint of increasing public virtue and interest in such matters, the campaign was an unqualified success."[22]

Operationally, a string of setbacks taxed Gray's patience. "Progress," Gray confided to Herms, "is disappointingly slow, but none the less sure." As always, funding was the chief problem. "The money available has been pitifully inadequate for the purpose. . . . Mosquito control," Gray explained, "has been the system of prophylaxis, this control being based on oiling of breeding places. No quinine prophylaxis has been attempted, both on the account of the lack of funds, and because of the American citizen's belief in his inherent right to poison himself and his family with patent remedies if he wants to."[23]

Oroville's health board had further complicated Gray's work when it reallocated a portion of the donations raised for mosquito work for a housefly reduction campaign. Even worse, the construction of a municipal sewer system during the summer created numerous cesspools that became major breeding areas for *Culex* mosquitoes.[24] Gray spent much of his time answering complaints from Oroville residents who blamed him for increasing the mosquito population.

In September 1911 Herms asked Gray to go to Los Molinos and prepare a mosquito survey for the land development company. Gray spent three days in Los Molinos. He reported to Herms that anopheline mosquitoes were breeding throughout the newly irrigated lands. He declared in his report, "I consider it is both possible and comparatively easy to eradicate malaria from this project with a reasonable expenditure of money, and within a reasonable length of time."[25] However, the Los Molinos Land Company could not solve the problem alone. "To get the most perfect and thorough results the company must obtain and maintain a cordial understanding and cooperation of the persons to whom the land is sold."[26]

Gray's mosquito survey had a positive effect. In 1912 the supervisors of Tehama County passed the "first sound county legislation in the State of California" against mosquitoes. County Ordinance 46 made it a crime punishable by arrest, fine, and imprisonment for any property owner to allow mosquito breeding to take place on his land.[27]

Encouraged by Tehama County's ordinance and the Los Molinos Land Company's endorsement of anti-mosquito work, Herms prepared a revised version of the Guill bill for the 1913 legislative session. New Jersey's 1912 mosquito commission law provided a model for the formation of county malaria control districts.

Reform-minded individuals were optimistic for the 1913 legislative session. California's legislature attracted national attention because of its progressive agenda. Suffragettes noted that California was one of ten states in which women could vote. "Since women got the ballot," Mary L. McClendon, president of the Georgia Women's Suffrage Association, declared, "[the California Assembly] has produced more advanced legislation desired by women than in the past twenty-five years."[28] McClendon observed that California women had lobbied for bills granting aid to needy parents, a health certificate law, a teacher's pension law, a state civil service law, and a bill for the creation of mosquito control districts.

Women from Penryn, Oroville, and Bakersfield campaigned for mosquito control and helped secure the state board of health's backing for Herms's revised version of the Guill bill. McClendon and other Progressives applauded the Assembly and the Senate's approval of the mosquito bill. Hiram Johnson, however, was unmoved by the women's appeals and vetoed the bill. The governor, Harold Gray acidly observed, believed that the mosquito control proposal "was merely a bill sponsored by a few medical crackpots for the benefit of a few farmers."[29]

The legislative session revealed unexpected support for the mosquito control bill in San Francisco. Malaria did not trouble the Bay area. But the salt marsh problem returned with vengeance in 1912. The citizens of Burlingame, San Mateo, and Hillsborough had failed to maintain Woodworth's and Quayle's early work. Ditches and dikes fell into disrepair. Two levees broke in the spring allowing waters to flow into the high marsh. A week later a spectacular flight of mosquitoes forced the owners of the posh Peninsula Hotel in San Mateo to close in the midst of the tourist season. By late spring "people who had to go out on the marsh," Harold Woodworth, Charles Woodworth's son, reported, "wore nets, tied ropes or strings around their ankles and wrists, wore gloves and even then were bitten." Conditions worsened in midsummer when "the mosquitoes migrated to town in a dark cloud for three days. Everybody who was not held in some way or another left town. It greatly affected property values and so it was seen that something radical must be done."[30]

Three South Bay executives resolved to resuscitate the Burlingame Improvement Association's fight against salt marsh mosquitoes. Nobel Stover, a civil engineer and one of Quayle's former students, contacted Charles Woodworth asking Woodworth if he would serve as the mosquito committee's advisor. Woodworth agreed. Stover, L. D. Whitney, and J. R. Fairbank raised $5,500 for anti-mosquito work. Whitney agreed to serve as the field superintendent.

Work began in February 1913. Woodworth and Herms helped Whitney recruit students from Berkeley and Stanford to serve as field inspectors. On weekends the students fanned out along the west side of San Francisco Bay looking for mosquito larvae. Whitney prepared flags and handbills for the students to post wherever they found mosquito larvae. The flyers read: "Please do not disturb this notice. It has been placed here by an authorized inspector of the San Mateo, Burlingame, and Hillsborough Mosquito Control."[31] On Mondays, teams of workers oiled the areas that the students had marked.

Whitney reported on the work's progress to the San Mateo, Burlingame, and Hillsborough Mosquito Committee in April 1913. "The extent of the infection," he wrote, "was almost beyond belief."[32] An early hatch of mosquitoes forced Whitney to devote part of his limited resources on oil. The mosquito committee was pleased with the results. "For the small sum of about $5,500," Harold Woodworth later observed, "these cities were freed of the pest which practically drove them from their homes the previous year."[33] Oiling was costly and provided only temporary relief. Banishing mosquitoes from the San Mateo and Marin counties required permanent improvements.

During the next winter Whitney and Stover "made great plans for a campaign for the whole bay side of the county."[34] Whitney divided the county into three districts (San Mateo, Redwood, and Menlo Park) at the beginning of the 1914 season. Once again students from Berkeley and Stanford served as part-time mosquito inspectors. All went well until May when "politics" forced the committee to stop work.[35] Whitney resigned. The salt marsh anti-mosquito work appeared to be finished.

William Herms was probably one of the few individuals who saw an opportunity in the committee's break-up. Herms's interest in salt marsh mosquitoes had grown since 1909. In 1912 the Smith Lumber Company in Contra Costa asked for Herms's guidance in reducing the local salt marsh mosquito problem.[36] He had noted that much of the support for the salt marsh work came "from real estate people who were having great difficulty in selling property because of the mosquitoes." After Whitney's resignation, Herms met with Stover and Harry T. Scott, a Hillsborough real estate tycoon. The three men settled on a plan of action. Herms and Stover set to work preparing a new mosquito bill for the 1915 legislative session. Scott promised to deliver the support of the real estate lobby in Sacramento.[37] Herms and Stover's bill passed both the Assembly and the Senate. On May 29, 1915, Hiram Johnson signed Assembly Bill 1540 into law. Six months later, Marin County earned the distinction of becoming the first mosquito abatement district (MAD). The Three-Cities District MAD (Burlingame, San Mateo, and Hillsborough) formed in mid-December. The mosquito abatement commission law, Harold Gray observed, was "signed by the Governor, who was uninterested in sickness and distress of rural residents, but who promptly paid attention to the screams of outraged realtors over lost profits."[38] Herms's four-year struggle to win passage of the California Mosquito Abatement Act exposed an underlying tension within the anti-mosquito movement. One faction called for the extermination of mosquitoes primarily for reasons of comfort. The other sought to wage war against mosquito-borne diseases. Herms's failures in 1911 and 1913 taught him a valuable lesson. If he were to succeed in his battle against malaria, then he would need to build a coalition between the different camps within the anti-mosquito movement. His eagerness to work with Nobel Stover and Harry Scott demonstrated that he had learned this lesson.

Anti-mosquito warriors in other states noted the political developments in New Jersey and California. In New York, residents of Long Island organized the Nassau–Suffolk County Mosquito Extermination Association at the beginning of 1916. Their objective was to "secure the passage of a bill authorizing the creation of a mosquito extermination commission for the county more or less on the lines of those in New Jersey."[39] New York City officials organized a "Mosquito Week" in which children distributed pamphlets calling for anti-mosquito work.[40] New York had adopted a mosquito commission law by year's end.

The campaign for relief from pest and nuisance mosquitoes stretched from New England to the Great Lakes. Citizen groups in Connecticut, Massachusetts, Pennsylvania, Delaware, and Illinois organized anti-mosquito leagues patterned on New Jersey's. Connecticut's agricultural experiment station published its first study of the state's mosquitoes in 1904. Seven years later progressives in Stamford organized a private subscription to drain two hundred acres of salt marsh. In September 1911 Wilton E. Britton, Connecticut's state entomologist, timed the release of a special bulletin, "The Mosquito Plague in Connecticut and How to Control It," to coincide with meeting of local health officers to discuss the need for legislative action against mosquitoes. Using New Jersey as their model, the meeting's participants mounted what proved to be a four-year drive to win an anti-mosquito law for Connecticut.

Victory came in 1915. Wilton Britton asked Spencer Miller and Thomas Headlee to come to Hartford to rally support for the anti-mosquito bill. "Today," Headlee told the Connecticut lawmakers, "New Jersey spends over $175,000 annually for mosquito control. . . . During the last seven years property along the shore has increased to over $5,000,000 in value." Headlee and Miller encouraged the legislators to follow New Jersey's lead. "A central controlling agency is needed," he explained, "because otherwise various municipalities will get careless and the work goes into discard."[41] The testimony proved persuasive. The bill passed the legislature, and the governor signed it into law. By 1916, the agricultural experiment station was coordinating the work of more than a dozen anti-mosquito leagues throughout the state.[42]

Baltimore, Philadelphia, and Chicago embraced the anti-mosquito movement during this period. In 1916 Baltimore spent $24,000 to fight pest and nuisance mosquitoes.[43] Philadelphia's effort to control mosquitoes began in 1912 when the mayor asked the city council to allocate five thousand dollars for mosquito control.[44] In the Midwest, the *Chicago Daily Tribune* inaugurated a citywide "Campaign against [the] pleasure killing pest" in July 1914. "The northern states, particularly those joining Illinois," the paper's medical correspondent explained, "have little to fear in the mosquito as a health underminer [*sic*]. The campaigns will be chiefly concerned in making the summer resorts, the camping places, the amusement parks and all the out of doors places for the greatest enjoyment with no drawbacks, such as the bites of mosquitoes."[45]

During the next two years, the *Chicago Daily Tribune* regularly published updates on the campaign. "Perhaps," a grateful reader wrote, "at this time a word on the mosquito question from Ravinia Park's standpoint will be in order. There has been a very decided improvement in conditions. I sat through the entire performance last night and the night before and was not molested."[46] Citizens in Winnetka, a prosperous North Shore suburb, followed the lead of South Orange and Burlingame and formed an improvement association in 1916. The *Daily Tribune* reported that the Winnetka Improvement Association had hired "the state bug expert" to survey the mosquito problem.[47]

Armed with the state's mosquito commission law, California's "bug expert" resumed his fight against malaria in 1916. In February 1916 Herms met in a closed session with the state's health officials and appealed for support for the fight against "the Minotaur of California." He told George Ebright, William Snow's successor as the health board's president, that a mosquito survey of all of California "was the most urgent necessity in the program of malaria control in the state."[48] A week later the California Health Board passed a resolution authorizing "in cooperation with the University of California, a survey of malaria and mosquitoes in California under the direction of Professor W. B. Herms, assisted by Mr. S. B. Freeborn."[49] The board allocated $2,150 for the first summer of the anticipated three-summer project. Herms noted that the $2,150 was restricted to covering "the cost of automobile, maintenance, hotel expenses, and general equipment, there being no charge made to the state board for the services of either the writer or Mr. Freeborn."[50]

Four days later, George Ebright used a meeting of the influential Commonwealth Club in San Francisco to announce the board's decision. Since its founding in 1903, the Commonwealth Club had championed progressive causes ranging from child labor laws (1906) to reducing air pollution (1913). Later Herms described this meeting as the "most important conference on malaria ever held in California."[51]

Ebright arranged for Herms; Karl Meyer, the new head of the George Hooper Institute; and Ray Lyman Wilbur, the president of Stanford University, to speak in support of the anti-malaria crusade. When the meeting ended, the Commonwealth Club had passed two resolutions: first, calling for the formation of mosquito control districts in "all malaria-infested areas in California . . . as soon as possible"; and, second, if this should prove impossible by the end of 1916, the state legislature should "appropriate funds to be used by the State Board of Health to employ sufficient numbers of inspectors to undertake the field work of malaria extermination."[52]

Herms and Freeborn's statewide mosquito survey was the first step in the board of health's campaign to eradicate malaria. The survey had three objectives. The first was epidemiological. Herms wanted to "ascertain the exact distribution of endemic malaria." Second, he wanted as an entomologist to "determine the distribution and occurrence of mosquitoes, particularly *Anophelines*" in California. Finally, Herms planned to use the survey and an educational tool "with [mosquito] control as the ultimate goal."[53]

On May 10, 1916, Herms and Stanley Freeborn began the first leg of an entomological odyssey that was to stretch across three years. A decade later Freeborn published the survey's findings in his *Mosquitoes of California*. Stanley Freeborn, who was born in 1891 and earned his bachelor of science degree from Massachusetts Agricultural College, had studied insect morphology and worked as deputy state nursery inspector before joining the Berkeley's entomology department as an instructor in 1914. Freeborn's already substantial skill as a taxonomist flourished in the first year's twelve-week expedition.

By August 14, Herms, Freeborn, and their student driver had traveled 7,036 miles in a "five passenger Chevrolet touring car . . . equipped for the summer's work in the field." A handful of Herms's graduate students in "a good old Ford" had accompanied the professors for the first six weeks of the survey. They had endured "rain, hail, snow, storm, heat, and cold . . . and visited the home of the mosquito and seen at first hand conditions good and bad as they actually existed."[54]

Most important of all, Herms rallied support for the formation of mosquito control districts wherever the Chevrolet stopped. In July the supervisors of Kern County approved the formation of the Dr. Morris Mosquito Abatement District (MAD). Named for a respected local physician who had led the fight against malaria, the Dr. Morris MAD was the state's first inland MAD formed to fight malaria. The Kern County trustees recruited Harold Woodworth, Charles Woodworth's son, to serve as the district's superintendent. By September when Herms returned to the lecture hall, citizen groups in Oroville and Pulgas had organized two additional inland mosquito abatement districts.

While Herms struggled against malaria in California, a pair of USPHS physicians and engineers were organizing a campaign against mosquito-borne

Figure 4.1. William Herms circa 1931. Photo courtesy of Alameda County Mosquito Abatement District.

diseases in the South. The formation of the USPHS out of the old federal Marine Hospital Service in 1912 marked a dramatic shift in the federal government's role in promoting the nation's health. To the USPHS's traditional responsibilities, Congress added a mandate that in the future the nation's health service was to "investigate the diseases of man and conditions influencing the propagation and spread thereof, including sanitation and sewage and pollution either directly or indirectly on the navigable streams and lakes of the United States."[55]

Buoyed by their success in Havana and Panama, veteran USPHS physicians Rudolph von Ezdorf and Henry Carter were "convinced of the necessity for, and feasibility of, malaria control in the United States."[56] Ezdorf, based at the Marine Service Hospital in Mobile, began collecting data on the incidence of malaria in the South in 1912. He ordered thousands of postcards sent to physicians throughout the region requesting information on the incidence of malaria in their communities. Simultaneously, Carter began to study the role of dams and hydroelectric impoundments in increasing the population of anopheline mosquitoes.

Woodrow Wilson's election as president in 1912 energized the progressive forces within the public health service. Three months after his inauguration, Wilson signed into law the Sundry Civil Appropriations Act authorizing $200,000 for the newly formed Division of Scientific Research of USPHS. In 1914 Rupert Blue, U.S. surgeon general, encouraged Ezdorf and Carter to ask Congress for "a few thousand dollars" for their anti-malaria work.[57] Ezdorf and Carter appeared before the Senate Appropriations Committee in February 1914. They persuaded the committee to earmark $16,000 for mosquito research and anti-malaria demonstration projects.[58]

In fact, Ezdorf and Carter had launched the USPHS's two "malaria demonstration" projects seven months before their Senate appearance in North Carolina. In 1913 Ezdorf received requests for help from the town of Fayetteville, North Carolina, and the owner of a textile company in Roanoke Rapids, North Carolina.[59] Roanoke Rapids and nearby towns of Rosemary and Patterson had a combined population of 4,100. Malaria was endemic among the mill workers. T. W. Long, the mill's physician, reported that nearly 75 percent of the population suffered from fevers and chills. Absenteeism periodically forced the mills to stop production. At the peak of the malaria season, July through October, Long treated an average of fifty malaria patients each day.[60]

Ezdorf and Carter agreed that Carter should go to Roanoke Rapids and meet with local officials. A four-day survey revealed that *Anopheles* larvae were present in the streams and run-off canals in the communities. Carter outlined to the company doctor, the mayor of Roanoke Rapids, and the mill's owners what was needed. The USPHS could not finance the work. If the mill's owners, however, provided the funds, Carter promised the USPHS's assistance in organizing the anti-mosquito campaign. The mill's owners pledged their support.

Carter, Ezdorf, and Joseph LePrince returned in October. Carter, who was both a physician and a civil engineer, performed a second anopheline survey. LePrince, who had spent ten years in Panama as William Gorgas's chief sanitary engineer, mapped out a drainage plan. Ezdorf, Long, and a nurse conducted a house-by-house malaria survey of the inhabitants of a four-block area in Roanoke Rapids. Analysis of blood smears from the five hundred people living in the survey area revealed that nearly half the population suffered from malaria.[61]

The anti-malaria campaign in Roanoke Rapids began in January 1914. Long lectured at local schools and before civic groups on the role of mosquitoes in transmitting malaria. Teachers and schoolchildren placed anti-malaria posters throughout the three communities. During the next six months Ezdorf, Carter, and LePrince regularly returned to Roanoke Rapids. Under LePrince's direction, workers dug close to seven miles of ditches, cleared forty acres of underbrush, and removed fifty-nine wagonloads of tin cans and other receptacles in which mosquitoes bred. By the end of the year company employees had used more than three thousand gallons of oil to kill mosquito larvae.[62]

The results were impressive. Long reported to Ezdorf that "during the summer of 1913, prior to any antimalarial work, the mills were constantly short of help on account of a large number sick from malaria. During the past summer there never has been a day when the mills did not have sufficient help." Blood smears revealed that the malaria rate had fallen to less than 5 percent of the population surveyed. A year later in 1915 Ezdorf performed a much larger malaria survey. His subsequent analysis of nearly one thousand blood smears drawn from schoolchildren revealed only thirty-four positives.[63]

Company officials in Roanoke Rapids were elated. "I will frankly admit," Samuel Patterson, treasurer and co-owner of the Roanoke Mill Company and Patterson Mill, told Ezdorf, "that I could not realize what a great change could be brought about by systematic work and with comparatively little expense. The money spent in antimalarial work here has paid the quickest and most enormous dividends that I have ever seen from any investment." To Patterson, the anti-malaria campaign was an essential component in his dream of making Roanoke Rapids a model for the new South. "I will close by adding that our experience has taught us that the eradication of mosquitoes is not only the proper thing to do from a strictly health standpoint but it is an exceedingly profitable thing to do."[64]

Carter presented an update on USPHS's anti-malaria work at the fourth annual meeting of the New Jersey Mosquito Extermination Association in 1917. Carter had become head of the USPHS's malaria research unit the previous year after Ezdorf's sudden death from a heart attack while on a return visit to North Carolina. Carter told the NJMEA delegates that the USPHS was guiding anti-malaria "experimental demonstrations" in North Carolina, Mississippi, Alabama, Missouri, Virginia, and Arkansas.[65] The USPHS work proved three things. First, malaria was prevalent throughout the South. Second, malaria annually exacted a tremendous cost both in millions of wasted lives and in tens of millions of dollars of lost production and medical costs. Finally, the USPHS teams had established that it was possible and economically feasible to eliminate malaria throughout the rural South.

The USPHS anti-malaria workers faced the same challenge that Herms encountered in Penryn. Who would pay for the projects? Not all mill owners shared Patterson's enlightened self-interest in Roanoke Rapids. Malaria was a poor person's disease concentrated in the rural backcountry. Local resources were insufficient.

The Rockefeller Foundation played a crucial role in improving health in the rural South. Stung by muckrakers' criticism, John D. Rockefeller pledged a million dollars in 1909 to a five-year campaign to eliminate hookworm in the South. Two years later Rockefeller sought a congressional charter for the Rockefeller Foundation. Critics howled. "This gigantic philanthropy," the *Los Angeles Record* declared, was nothing more than an attempt "by which the old Rockefeller expects to squeeze himself, his son, his tall-fed collegians and their camels, loaded with tainted money, through 'the eye of the needle.'"[66]

Congress declined the billionaire's request. New York's lawmakers, however, issued a charter in 1913. Three years later Wycliffe Rose, director of the Rockefeller Foundation, contacted the USPHS and asked for recommendations for future health projects by the foundation's International Health Board (IHB). In 1916 the IHB began what was to be a thirty-year effort to eradicate malaria in the South and throughout the world.[67]

The IHB and the USPHS's first joint initiative was an experiment in 1916 designed to weigh the effectiveness of two anti-malaria strategies. Carter chose two small towns in southeastern Arkansas for the test. In the first, Crossett, USPHS workers employed anti-mosquito measures such as ditching and oil to eliminate *Anopheles* mosquitoes. In the second, Lake Village, IHB funds paid for screening and "immunizing doses" of quinine for the residents. The anti-mosquito work in Crossett led to a 77.33 percent reduction in the town's malaria index during the height of the disease's "active" season as compared to a 70.6 percent reduction in the parasite index in rural residents living near Lake Village. The cost per resident of the anti-mosquito work was $1.23 as compared to $1.75 for quinine and screening. The Arkansas IHB funded project revealed that both strategies were effective in combating malaria. Anti-mosquito measures, however, were less expensive while offering a greater reduction in malaria.[68]

The Arkansas work exposed significant disparities in the living conditions of poor whites and African Americans. USPHS and IHB inspectors reported that 98.3 percent of the windows in the white section of Crossett were screened as compared to 70.8 percent of "colored" homes.[69] In rural Lake City, Robert Derivaux, the lead USPHS investigator, discovered that in the poorest households "both whites and colored . . . [were] almost totally unprotected and represents the class of which the ravages of chronic malaria are more prominently manifested."[70]

Racism compounded African Americans' exposure to mosquito bites and malaria in three ways. First, USPHS and IHB researchers readily acknowledged significant discrepancies in their surveys of rural white and black populations. "Census taking in colored sections," USPHS sanitary engineer A. W. Fuchs reported, "we found more difficult than among whites. Out of fifty homes visited in Little Washington section of Goldsboro [North Carolina], only two positive histories were obtained. Yet one Negro doctor reported three hundred and fifty cases of malaria in Little Washington during 1919." Fuchs justified the IHB and USPHS's failure to measure the malaria rate among rural African Americans in blatantly racist terms. "The darky is by nature secretive regarding such matters and entertains a fear of quarantine or vaccination in spite of all explanations to the contrary."[71]

IHB sanitary engineer C. E. Buck explained the underreporting of African American malaria rates in less prejudicial language. Buck identified two problems with malaria surveys in black communities. First, local organizations "such as the Women's Club, civic league, or high school students" typically overestimate the malaria rate in the white community "because they are anxious to have the work done, and in their zeal to have the program go through exaggerate the amount of malaria that exists." The opposite happens in the black community. Civic groups consistently "fail to interview many of the Negroes, who usually live on the outer edges of town, where malaria is most prevalent, with the result that in this respect the figure is too low." Employing "trained men, usually outsiders" to perform the survey only compounds the problem in underreporting because "people are loath to admit to outsiders that they have had malaria. Negroes are particularly prone to say no to any questions regarding malaria for fear."[72]

Underreporting the malaria rate in the African American community was widespread in the years immediately before and after World War I. Protecting rural African Americans from mosquito-borne diseases was not a health priority. Senior USPHS, IHB, and state health workers defined their objective as eliminating malaria in the white population. "The victory of science," Bruce Mayne, a USPHS associate sanitarian, explained "in drawing the teeth of the vicious mosquito has blazed the way for scientific sanitation and health, and has made possible such achievements as the domination of the white man in the Tropics."[73]

Soon, USPHS sanitarian L. M. Fisher predicted, southern politicians would realize that malaria was a "problem of major importance to state boards of health in the South." Recognition of this fact was the first step toward "reclaiming the most fertile lands in the country and making them habitable for the white race."[74] James Hayne, South Carolina's health officer, agreed stating his agency's anti-malaria goal in unalloyed racial terms. "The white man will not live in a country where there is malaria. . . . The reason of the black belt is malaria. When you remove that black belt then you wipe out the blackening of that part of the country. The Negro will live where he has malaria, the white man will not."[75] For Hayne the anti-mosquito campaign was ultimately about dollars and cents. "South Carolina is a little bit of a state. . . . The Negroes don't pay taxes. Get that in your head when you talk about the amount of appropriation."[76]

There were dissenters to Hayne's racist perspective. "We are fighting malaria for humanity," said W. H. De Mott, who led anti-mosquito work in the newly formed Nassau County, Long Island, abatement district. "It doesn't make any difference whether it is a negro or who it may be."[77] De Mott, however, was in the minority.

Fortunately, the research of USPHS entomologists on anopheline mosquitoes was not tainted with the sanitarians' racism. Two years prior to the Crossett and Lake Village study, Bruce Mayne, Joseph LePrince, and Thomas H. D. Griffitts had conducted a series of experiments that held far-reaching implications for the war against the mosquitoes. Beginning in 1914 in Scott, Mississippi, and later in Monroe, Louisiana, Bruce Mayne compared the role of different species of anopheline mosquitoes in transmitting malaria to human hosts. Today, entomologists know that there at least four hundred species of *Anopheles* mosquitoes throughout the world. Roughly seventy of these species are competent vectors of the malaria plasmodia. The challenge was identifying which of these species posed the greatest health threat.

In the Southeast, at least thirteen different species of *Anopheles* mosquitoes are potentially vectors for malaria.[78] Three of these species are particularly troublesome. *Anopheles atropos*, a brownish-black mosquito which breeds in brackish water, is a fierce biter but is not known as a vector for malaria. *Anopheles crucians*, distinguished by pale scales on its wings, is a competent vector for malaria that prefers to feed on livestock and other large mammals.

Mayne's experiments revealed that the third of these species, *Anopheles quadrimaculatus*, was the principal vector of malaria to humans in the South. Taxonomists now consider this mosquito a species complex comprised of five different but similar mosquitoes. *An. quadrimaculatus* is a freshwater breeder, feeds during the night, and is characterized by "having four, more or less distinctive spots on the wings."[79] Mayne's discovery meant that sanitarians could develop programs of "species sanitation" similar to Watson's work in the Malay States. It was not necessary to exterminate all anopheline mosquitoes. Control of a single species offered the possibility of eliminating malaria.

LePrince and Griffitts's work was equally revolutionary. In 1914 LePrince and Griffitts conducted the first flight range experiments for anopheline mosquitoes in Fort Lawn and North Augusta, South Carolina. LePrince and Griffitts applied aniline dye to adult *Anopheles quadrimaculatus* and released them. Collection traps were placed at specified distances from the release point. LePrince and Griffitts demonstrated that the mosquitoes' flight range was limited to between one-half and one mile from the release point.[80]

Mayne, LePrince, and Griffitts's discoveries came at a critical moment. The United States entry into World War I sparked a dramatic expansion in the mosquito control movement. On April 3, 1917, the day before Congress declared war on the German Empire, Woodrow Wilson issued an executive order incorporating the USPHS into the United States military forces and charging it with responsibility for care "of sick and wounded officers and men or for such other purposes as shall promote the public interest in connection with military operations."[81] In the history of war, insect-borne diseases had exacted a tremendous toll in human fatalities. The Spanish American War revealed the threat that such insect-borne diseases as yellow fever posed. In what Secretary of State John Hays described as a "splendid little war," only 862 men perished on the battlefield while 5,438 died of disease.[82]

During the twelve months following the outbreak of hostilities, American military forces grew from two hundred thousand to over two million men in the uniformed services. The army and navy rushed to open new training camps and recruitment centers. Many of the military facilities were located in areas in which malaria was endemic. The War Department divided responsibility between the USPHS and the army and navy for the nationwide anti-mosquito initiative. Army and navy physicians were in charge of anti-mosquito work within military reservations. War Department officials assigned the USPHS the task of organizing anti-mosquito work in civilian areas bordering military bases. By the war's end, USPHS sanitary engineers, entomologists, and physicians had established anti-mosquito programs at forty military facilities in fifteen states. At its peak, this work protected nearly five million civilians and military personnel.[83]

Joseph LePrince, Henry Carter, and Robert Derivaux guided the USPHS wartime mosquito and disease control work. Derivaux, who had directed IHB Arkansas demonstration work, organized anti-mosquito programs in southern Mississippi and Louisiana. Carter coordinated the extracantonment programs in Maryland and Virginia. LePrince took charge of the remaining southern and eastern states.[84]

The USPHS extracantonment work stimulated the growth of mosquito control throughout the South. Florida offers a prime example of this. The Sunshine State's war against mosquitoes grew with the USPHS's assistance into a powerful, statewide movement during World War I. Cordial relations had long existed between Joseph Porter, Florida's chief health officer, and the USPHS

Division of Scientific Research. The division's mailings of malaria survey cards provided critical data to Porter's newly formed Bureau of Vital Statistics.

LePrince dispatched Calvin Harrub, a USPHS sanitary engineer, to Florida in 1918 to organize mosquito control efforts at the U.S. government military reservation in Jacksonville. Harrub asked Porter's advice on who was best qualified to guide the campaign. At Porter's suggestion, Harrub requested that George Simons, who led the Bureau of Engineering at the Florida Board of Health, to make a sanitary survey to "remedy and eliminate nuisances of potentially dangerous character."[85]

Army officials feared that recruits would contract malaria from the local population. Harrub and Simons developed a detailed plan to protect the military installation from mosquito-borne diseases. They proposed a twofold strategy. On base, the military would enforce a stringent sanitation regimen. Barracks would be screened, trash and rubbish cleared, and any pools of standing water drained or treated with oil or kerosene. The greater threat came from outside the base from *An. quadrimaculatus* mosquitoes that abound in the flat piney woods of north Florida.

Mayne's research demonstrated the importance of preventing *An. quadrimaculatus* from reaching the military encampment. LePrince and Griffitts's flight range experiments guided Simons's Jacksonville anti-malaria work. Since the flight range of *An. quadrimaculatus* mosquito was seldom greater that one mile, Simons and Harrub prepared a detailed plan of drainage ditches that guaranteed the run-off of water and prevention of mosquito breeding within the critical zone.

Porter approved Harrub and Simons's program of ditching and drainage. Until World War I the thrust of anti-mosquito measures aimed at preventing insect bites. Public health officers at Camp Johnson sought to eliminate the problem's source. "No where in Florida," the *Florida Health Notes* reported, "has the importance of Malaria Control Work been realized as much as in the territory surrounding the U.S. Government Reservation at Jacksonville."[86]

The wartime anti-mosquito effort spanned the continent. Along the eastern seaboard and in the Mississippi Delta, LePrince recruited members of the New Jersey Mosquito Extermination Association to supervise the extra-cantonment work. LePrince made John Smith's 1904 call for "trench warfare" against mosquitoes the marching orders for the newly commissioned sanitary officers. Local authorities in Bergen County learned in the fall of 1917 of the War Department's plan to establish an embarkation camp that would house thirty thousand soldiers in an area that was notorious as a "habitual breeder of malarial mosquitoes." LePrince asked the Bergen County Mosquito Extermination Commission to take responsibility for "strenuous efforts to make Camp

Merritt mosquito-proof." Guided by John Smith's and Thomas Headlee's work, USPHS anti-mosquito physicians and engineers succeeded in protecting tens of thousands of soldiers at Camp Merritt from malaria.[87]

When the Armistice was declared, members of the NJMEA led anti-malaria programs that stretched from Camp Dix in New Jersey to Camp Pike outside Little Rock, Arkansas.[88] The commanding officer at Camp Pike offered a special commendation to Capt. Russell Gies, formerly Union County's (New Jersey) Mosquito Commission's superintendent, for his work to suppress malaria. "Captain Gies," the *Arkansas Democrat* reported, "has had much experience combating the festive mosquito, for he hails from New Jersey where he was employed by the State previous to the war in a similar line of work." Gies and his assistant and fellow Jersey man, Walter Talbot, demonstrated that "freedom from mosquitoes is attained only at the expense of eternal vigilance."[89]

World War I had far-reaching consequences for mosquito control in the Pacific West. In California, the war forced Herms and Freeborn to rethink their plans for the second year of the statewide mosquito survey. Freeborn was unable to participate. State officials assigned him the task of establishing mosquito control for wartime labor camps. On May 21, 1917, Herms headed south from Berkeley along the coast down to San Diego to continue the mosquito survey in southern California. Herms spent much of the summer guiding "intensive work in the vicinity of military cantonments and other smaller military camps."[90]

Herms volunteered for active duty in the army when he completed the California extracantonment work. His first assignment took him to Texas. In October Captain Herms received orders to report to Newport News, Virginia, and assume responsibility for organizing anti-malaria work at camps Stuart, Hill, and Alexander. Henry Carter and Thomas Griffitts had conducted a malaria survey of this area in August. Their unpublished report revealed that the three camps were located in the middle of a malarial swamp. The high number *An. quadrimaculatus* mosqitoes and the endemic presence of malaria in the local population posed a tremendous threat to the thousands of inductees preparing for embarkation. "I regard," Col. J. B. Clayton wrote to Surgeon General William Gorgas, "the proper drainage of these swamps as the most important sanitary work we have to do in this vicinity, as, if it is not done properly, troops stopping here while waiting for overseas transportation . . . will unquestionably become infected with malaria in spite of netting and mosquito bars."[91]

Virginia was infamous for its malaria problem. Mark Boyd, who led the IHB's Malaria Treatment Station in Tallahassee, contended that "disease nearly caused the failure of the Jamestown settlement of the London Company" in

1607.[92] Conditions along the Tidewater Basin had not significantly improved in the intervening three hundred years. Henry Carter noted during a USPHS survey of Wilson, Virginia, in 1915 that "every house I visited in early October had a sick inmate, and in some there were several."[93] William Schoene, Virginia's state entomologist, extended Carter's observations to the entire state. "The mosquito problem in Virginia," Schoene told delegates to the 1917 NJMEA annual meeting eight weeks before America's entry into the war, "is almost untouched considering the fact that there are a million acres of swamp and marsh land in the State, this is a virgin field for eradication work."[94]

Herms faced two immediate challenges when he arrived in Newport News. First, he needed a detailed survey of the local anopheline situation. Second, a successful anti-malaria campaign required skilled mosquito inspectors and sanitary workers. Herms took charge of the mosquito survey. A growing labor shortage meant that Herms had to recruit and train his own workforce. Herms persuaded the camp's surgeon general to assign two hundred inexperienced African American inductees to his command. During the next six months, Herms conducted a malaria boot camp in which the men learned the basics of mosquito control. Herms received a late Christmas gift in December when Stan Freeborn arrived. Freeborn had volunteered for military service and convinced army officials to assign him to Newport News as Herms's assistant.

On April 8, 1918, Herms and Freeborn's sanitary unit was reorganized as the Malarial Drainage Detachment. Ditching commenced in mid-April. The situation was dire. "Briefly, the camps constituting this port of embarkation are in swamps," Herms's commander confided in a letter to his superiors, "or so near swamps that to all intents and purposes, so far as the mosquito menace is concerned, they might as well be in the swamps. Mosquitoes breed by the million in the swamps in question. These mosquitoes are of the malaria carrying variety."[95] The well-being of tens of thousands of soldiers and sailors depended on the success of Herms's anti-mosquito detachment.

The results were astonishing. Between the formation of the Malarial Drainage Detachment on April 8 and the year's end, army physicians treated 144 cases of malaria at the embarkation hospital.[96] Herms's experiences in the Sierra country served him well. An examination of the patients' case histories revealed that not a single case of malaria had originated in the area under his command.[97]

World War I demonstrated the value of mosquito control and medical entomology. Three and one half million Americans served in the armed forces. Military records showed that there had been 1,400 cases of malaria with only 31 deaths attributed to the disease. "Had the same [malaria] rate prevailed as in the Spanish-American War," the army's surgeon general estimated, "there would have been 1,900,000 cases of malaria and 5,600 deaths" during the eighteen months of war.[98]

The war's end meant that Herms and Freeborn could return to California and resume their effort to protect civilians from mosquito-borne diseases. Charles Woodworth announced his retirement in 1919. University officials chose Herms to head the newly created Division of Entomology and Parasitology. In March 1920 Herms organized a two-day meeting at Berkeley to consider the mosquito control's future in California. Informal get-togethers at the faculty club continued throughout the decade. Herms recommended in 1930 that the participants follow New Jersey's and Florida's lead and form the California Mosquito and Control Association (CMCA). The participants agreed, naming Nobel Stover the CMCA's first president and Harold Gray the association's secretary.

Like Headlee, LePrince, Carter, Porter, and Simons, Herms's wartime experiences convinced him that the mosquito control movement was destined to grow in the coming decade. Given the resources, the mosquito crusaders could vanquish their ancient foe. "God speed the day when their seed shall no longer blight this fair state," Herms prophesized in an impassioned speech before the California Board of Health. Echoing the cadences of an old Christian hymn, he declared, "The fight is well begun, the battle is on, and victory will certainly follow."[99]

Chapter 5 Widening the Campaign

In 1921 the British writer D. H. Lawrence, with the experience of pest-filled Venetian nights still fresh, penned a seventy-four-line, free verse poem titled "The Mosquito." Lawrence, whose explicit descriptions of human sexuality seven years later in *Lady Chatterley's Lover* shocked his contemporaries, proved less competent in matters concerning the male's role in nature's great drama of blood and sex.[1] Six years later William Faulkner chose *Mosquitoes* as the title for his second novel. Drawing on a seven-month stay in New Orleans, Faulkner used a boating excursion on Lake Pontchartrain as the frame for his mocking portrait of the Crescent City's artistic and socially prominent pests. A flight of salt marsh mosquitoes set the novel in motion. "They came cityward," Faulkner observed in the novel's dedication, "lustful as country-boys, as passionately integral as a college football squad, pervading and monstrous but without majesty: a biblical plague. . . . The majesty of Fate become contemptuous through ubiquity and sheer repetition."[2]

While Lawrence and Faulkner mused on insects, organized mosquito control made significant advances in reducing mosquitoes' injury to the body politic. Encouraged by their wartime achievements, advocates of mosquito control renewed their efforts to win local and state support for anti-mosquito measures in the 1920s. Utah, Florida, and Illinois had passed laws allowing the formation of mosquito abatement districts by 1930. Simultaneously, scientists in Louisiana and New Jersey developed chemical agents opening new possibilities for control.

Herms and Freeborn returned to Berkeley in early 1919. Their success in Newport News made them eager to resume anti-malaria work in California. "California," Herms wrote to his former colleagues in the U.S. Public Health Service (USPHS) shortly after his return, "has made remarkable strides in the control of malaria in the past 10 years, having reduced the prevalence of this disease by at least 60%."[3] The dark shadow of malaria, however, continued to waste the lives of thousands of farmers living in the Sacramento and San Joaquin valleys. In October 1919 Herms told the delegates attending the annual conference of state, county, and municipal health workers that "the control of malaria presents the biggest rural sanitary problem in California today."[4]

Herms's immediate objectives were to complete the statewide 1916–1917 anopheline survey, expand the number of mosquito abatement districts, and reorganize the university's entomology department. Herms planned to restart the survey in June. Early in the spring, Harold Gray, Herms's former student, sent word of an alarming outbreak of malaria in northern California. This epidemic gave Herms and Stanley Freeborn an unforeseen opportunity to demonstrate the effectiveness of anti-mosquito skills they employed in Virginia.

Harold Gray provided a detailed description of the growing crisis. Gray had served as a district health officer for the California Department of Public Health since 1917. Having abandoned his plan to earn a doctorate in 1912 after his wife gave birth to their first child, Gray served as the municipal health officer in Palo Alto and San Jose between 1912 and 1916. At the beginning of World War I, Gray placed fourth, highest among nonmedical candidates, in a USPHS nationwide examination. This earned him the position with the California Department of Public Health. Gray's territory covered seventeen counties in northern California where his duties included oversight of water and sewage facilities, stream pollution, and malaria control.

From his base in Chico, Gray was able to monitor the growing malaria problem in the Anderson Valley section of Shasta County. Malaria was long endemic in the region. The focal point of Gray's concern was the small, unincorporated town of Anderson. Located twelve miles south of Redding, the Anderson Valley had a population of approximately 1,300, of which roughly 450 were concentrated in the town of Anderson. The remaining population were farmers (650) or resided in nearby Cottonwood (200).[5]

Conditions in the Anderson Valley worsened after 1918 with the completion of the Anderson-Cottonwood Irrigation District project. Seepage from the drainage district's canals, failure to maintain lateral ditches, and the introduction of rice cultivation created numerous standing pools of water throughout the district that were "ideal for the breeding of anopheline mosquitoes." Gray feared that the poorly designed irrigation project would exacerbate the endemic malaria problem. In 1918 he visited virtually every household in valley. "Practically every one" Gray encountered "in the irrigated sections complained bitterly about the mosquitoes, and in many places their presence seems to have made life nearly unendurable."[6]

Wilbur A. Sawyer, the state's health officer, carried Gray's disturbing report to the governor's office. William Stephens, a Los Angeles Republican who had earned a fortune as a wholesale grocer, knew from personal experience the devastating effect that malaria had on rural farming communities. Governor Stephens ordered an emergency $10,000 appropriation to the state board of health. Additionally, he directed Sawyer to establish a mosquito abatement district in Anderson.[7]

The Anderson Mosquito Abatement District was unique. This was the first time that the state launched a mosquito control program. Chapter 584 limited the taxing authority of the mosquito abatement commissions. Districts could not exceed a ten-cent tax cap, and Anderson was too small to mount an anti-malaria campaign. In 1922, the first year after the approval of an abatement tax, Anderson generated $380 for mosquito control. It would have taken the city more than twenty-five years to raise $10,000 for the needed anti-malaria work. This was why Stephens offered the emergency relief.[8]

Fear that Chinese and Japanese immigrants would replace Anderson's white population supplied an additional incentive. At least, this is Linda Nash's argument in *Inescapable Ecologies*. "It is perhaps not coincidental," Nash noted, "that California's first state-funded malaria control effort occurred in Anderson, a town that public health investigators took note of for its 'strikingly pure population of native born Americans.' The attempt to eradicate malaria in the Central Valley would be merely a new chapter in the long-standing effort to ensure that California would become and remain the home of those who called themselves 'white.'"[9]

Racism is a recurring motif in the history of malaria. It was certainly present in California. The primary motivation for Sawyer, Herms, and Freeborn, however, lay in the skyrocketing increase in malaria that followed the opening of the Anderson-Cottonwood irrigation district in 1918. The numbers were staggering. By 1919, 54.5 percent of the population within the irrigation district's boundaries suffered from malaria.[10]

Sawyer asked Herms and Freeborn to help Anderson. Gray would serve as their field representative. Before the plans were finalized, Sawyer left California to accept a position with the Rockefeller Foundation's International Health Board. His successor sought to reduce district health officers' independence. Gray had forcefully advocated "developing programs on a direct local basis." His new supervisor disagreed. "H.F.G. was young," Gray wrote of himself thirty-five years later, "and to say the least, not too tactful. He stepped on big political corns repeatedly. . . . With mutual pressure, he was fired on July 1, 1919."[11]

Stan Freeborn took charge of the Anderson work. Freeborn made his first visit to Anderson in June shortly before Gray's dismissal. The first order of business was to determine the town's malaria rate. Later, Freeborn acknowledged that when he arrived in Anderson he was skeptical of the "extravagant hearsay evidence regarding the prevalence of the disease that had been sifting to the administrative offices." Freeborn set up "temporary laboratories on street corners, in stores, and private homes" and secured 119 blood smears. A quarter of the adults tested positive. Nearly 36 percent of the children and teenagers between the ages of ten and fifteen carried the parasite. The actual infection rate was probably higher. Subsequent interviews with the sample population revealed that forty-three of the individuals whose blood smears tested negative were already taking "chill tonics" and quinine. Freeborn's street corner collections suggested that between 60 and 100 percent of Anderson's residents suffered from malaria.[12]

While the town's doctor and his nurse catalogued the blood smears, Freeborn made a hurried survey of the town's mosquito breeding sites. His observations confirmed Gray's findings. Run-off and leakage from poorly planned irrigation ditches kept low-lying areas inundated long after the spring rains ended. Anderson abounded in sites producing anopheline larvae.

Freeborn prepared a detailed malaria eradication plan for Anderson. There were three components in Freeborn's proposal: permanent reduction of the mosquito population, treatment of active malaria cases, and an aggressive effort to screen homes and buildings in the community. During the next six months, workers dug close to eighty thousand feet of ditches that provided drainage for the poorly planned irrigation work. By year's end, the town's physician and nurse had treated five hundred people for malaria. To prevent the spread of malaria the next spring, Freeborn had organized a "screening epidemic" in Anderson. If local residents would buy screens, Freeborn promised to pay for the installation.[13]

The Anderson anti-malaria work was a modest success. In December Freeborn conducted a second malaria survey. He collected seventy-six blood smears of which sixty-four (84 percent) were negative. Analysis of the twelve positive samples revealed that in only five cases was the malaria plasmodium in its active phase. "The final results of this entire project," Freeborn wrote in his report to the state health officials, "rest in the future as the next mosquito season will demonstrate the success or failure of the measures undertaken." Regrettably, Freeborn confided to Herms, the ten thousand dollars was exhausted before the anti-mosquito work had reached "an ideal point."[14]

Anderson's residents embraced mosquito control. In 1920 the editor of the *Anderson Valley News* observed that "the presence of a single mosquito is *NOW* [emphasis in original] a call to arms and the occurrence of an isolated case of malaria calls forth the wonder and sympathy of the entire community."[15]

Herms congratulated Freeborn on the Anderson campaign, arguing that it was a model for future state anti-mosquito work. "What Anderson has accomplished," Herms declared, "other California communities can accomplish, providing funds are available for fundamental work, the cost of which is beyond the means of the community to assume. . . . Happy and grateful Anderson wants the state to know of its victory, its prosperity, and its gratitude."[16]

In 1923 Rockefeller Foundation's IHB sent Louva Lenert, a sanitary engineer, to review California's progress in mosquito control. Lenert and Edward Ross, chief sanitary inspector for the CBOH, visited Anderson while preparing a detailed analysis of the status of malaria control in California. Since the passage of the 1915 abatement law, fifteen communities had organized mosquito abatement districts (MAD). Ten districts were located in the Sacramento and San Joaquin valleys. Nevertheless, the Central Valley continued to have a malaria problem.

A pattern developed throughout the valley. As the number of irrigated acres rose from 1.4 to 4.2 million acres between 1900 and 1920, the immigrant laborers sought jobs on the newly created farmlands.[17] Herms maintained that "man's own carelessness" had also produced ideal larval habitats for disease-bearing mosquitoes.[18] As was the case in Anderson, none of the MADs possessed sufficient resources to mount effective anti-malaria campaigns. "The unfortunate feature," Lenert and Ross observed, "is that property values are not high enough to provide sufficient funds with the ten cent limit of taxation."[19]

High among Herms's worries was the tremendous growth in rice cultivation in the Central Valley. Rice was introduced as a commercial crop in 1912. By 1916, eighty thousand acres were annually flooded to allow rice cultivation. "The culture of rice," Herms observed, " . . . usually if not always increases more or less considerably the number of mosquitoes and the cases of malaria." Herms feared that without mosquito control it would become a "menace to public health." Four years later rice cultivation exceeded 160,000 acres.[20]

Rice farming was only part of the challenge facing mosquito control in California's inland districts. Lenert and Ross's visits to nine of the ten districts revealed a shocking "lack of knowledge of the latest known control methods." The Sacramento and San Joaquin valley districts operated on their "own initiative and without advice as to the most effective and economical methods to be applied."[21] Eliminating malaria was essential for the economic development of California's interior. "The indelible stamp of this disease [malaria]," Lenert and Ross concluded, "is apparent in various districts. . . . Its elimination is dependent upon the foresightedness of those concerned with the greatest possible development."[22]

While Freeborn was organizing the Anderson program, Herms completed the survey of the state's anopheline mosquitoes and began his effort to reform the entomology department. By 1920 Herms had traveled more than 18,000 miles, crisscrossing the state. He made more than 690 mosquito collections in the course of visiting every county in the state. Herms assigned Freeborn the task of cataloguing the 6,650 mosquito samples amassed during their summer expeditions. Using this as the basis for his dissertation, Freeborn received his Ph.D. from the University of Massachusetts in 1924; his *Mosquitoes of California* appeared two years later.[23]

Three thousand miles east of Berkeley, Joseph LePrince faced even more frustrating challenges in his effort to expand mosquito control in the South after World War I. Born in 1875 in Leeds, England, LePrince and his parents immigrated to New York when he was twelve years old. In 1898 he earned a degree in civil engineering from Columbia University. Six years later William Gorgas recruited LePrince to serve as his sanitary engineering assistant in Panama. LePrince's zeal for the mosquito crusade was remarkable. Leland Howard retained a clear, if bemused, memory of their first encounter. Colonel Gorgas with LePrince visited the Bureau of Entomology to get Howard's advice. When the colonel and his aide stood to leave, Howard asked a favor. It would be a great help if Gorgas would send specimens of Panamanian mosquitoes back to the Bureau. "I will assign Mr. LePrince," Gorgas told Howard, "to see that this is done." LePrince nodded his assent adding, "I will have to do it soon, Doctor, for in a year or so there will be no mosquitoes there!"[24]

Ten years in Panama and his experiences leading the USPHS's wartime extracantonment work on the East Coast had not dampened LePrince's passion for the anti-mosquito cause. After the war, LePrince renewed his campaign against malaria in the South. LePrince used his position as director of the USPHS's postwar Malaria Investigations Division to encourage municipal and county anti-mosquito initiatives.

Chatham County, Georgia, was one of LePrince's first successes. In 1920 workers began repairing tide gates and digging ditches in a twenty-five-square-mile area surrounding Savannah to provide drainage for abandoned rice fields. "The malarial incidence," Leland Howard reported, "was reduced enormously by these operations."[25] By 1922 the USPHS's Malaria Investigations Division had supported anti-mosquito demonstration projects in eleven southern states. The total appropriation was sixty thousand dollars.[26]

Money remained the principal obstacle in advancing the anti-malaria campaign. Congress slashed the USPHS budget after World War I. During the war, Washington allocated a million dollars for anti-mosquito work. Lunsford Fricks, a veteran USPHS surgeon, told delegates to the USPHS's Second

Annual Antimalaria Conference, that the "demonstrations in malaria control were early recognized as one of the most fruitful fields of investigation . . . and that there is still a legitimate need for malaria-control demonstrations." However, the USPHS lacked the funds to support the work. James Hayne, South Carolina's health officer, feared that the anti-malaria campaign had stalled. "If we are going to do anything in 1921," Hayne declared, "it will be to hold what we have and not do very much more."[27]

Even a small amount of money could make a substantial difference. This was certainly the case in Florida where LePrince's approval of a USPHS anti-malaria demonstration project proved indispensable. Late in 1919 George Simons, Florida's sanitary engineer, received a request from the Burton-Swartz Cypress Company and the city council in Perry, Florida, for emergency assistance. Located in the flat piney woods and cypress swamps of the Florida Panhandle, Perry had one of the highest malaria rates in the South. Collectively, Perry's pharmacists estimated that the town's two thousand citizens annually spent three thousand dollars for quinine and "chill tonics." Local physicians reported that 50 percent of the people suffered from malaria.[28]

Perry provided Simons with an opportunity to demonstrate the effectiveness of mosquito control. Simons contacted his wartime colleagues at the USPHS and asked for their assistance. LePrince sent J. G. Foster to help Simons in drafting his plan. Simultaneously, Foster and Simons rallied support for this "progressive undertaking" among town's business and political elite.

It took six months to raise the funds. The town contributed fifteen thousand dollars to the project, and the owners of the Burton-Swartz Cypress Company contributed ten thousand dollars. Cash-strapped county officials added three thousand dollars. Work began in May. By the project's completion in November, Simon's team had opened three miles of creek beds, cleared fifteen acres of land, dynamited an old dam, and constructed three new bridges.[29] The Perry project transformed mosquito control in Florida. The project dwarfed earlier undertakings. George Simons described the Perry Project as "one of the largest of its kind in this country."[30] One year later, the malaria infection rate was reduced by 90 percent.[31]

Mosquito control revitalized Perry. The owner of the Burton-Swartz Cypress Company wrote Simons and described the positive consequences of mosquito control. "Before your work here," E. G. Swartz wrote, "there were times in the late summer and autumn months that our business was seriously handicapped by being forced to operate our plant shorthanded, due entirely to malaria among our employees." J. H. Scales, a local banker and member of the Florida Senate, echoed Swartz's enthusiastic appraisal. "Doctors tell me that they have less than one-tenth the malaria and chill patients than three years ago."[32]

Simons's crusade for mosquito control was not restricted to Perry. He found an unexpected ally in a reform-minded physician one hundred miles to the south in the phosphate mining lands of Polk County in Brewster, Florida. Depending on the year, between 40 and 80 percent of the fifteen hundred residents suffered from malaria. At times, malaria kept one hundred workers out of the mines.[33] William Buck, Brewster's sole physician, persuaded the American Cyanamid Company to support a public health initiative.

Buck sought to counter the prejudice against poor white and Negro populations. Poor whites in the South, called "crackers" in Florida, were thought to be by nature "defective in cleanliness." Contemporary accounts described crackers as living in "heaps of filth" possessing "absolutely no virtues, and . . . dirtier, if possible than the Negro." Adults bore a "pleading wistful look about the eyes (which seem to have lost whatever light they once had)." Cracker children were "almost uniformly pot-bellied."[34]

Buck believed that an improved diet, sanitation, and eliminating malaria would transform Brewster. The mosquito problem was the most vexing challenge. Ditching alone proved ineffectual. "The first few weeks of our work demonstrated," Buck wrote in a 1920 report, "the impossibility of eliminating the mosquito by drainage and oiling alone; however, close and repeated inspection failed to reveal mosquito larvae in several of the larger ponds and several of the slow moving weed-filled streams, the natural habitat of the *Anopheles*."[35] Buck discovered that the ponds and streams, which were free of mosquitoes, contained minnows. The fish, *Gambusia affinis*, consumed the mosquito larvae before they could pupate.

With too much land to drain, fill, or oil, Buck hit on the idea of basing the campaign around minnows. His first step was to establish a minnow-breeding pond. Once assured that he had a sufficient supply of minnows, he outfitted what was probably the first mosquito control vehicle in Florida history. In 1920 Buck's "mosquito wagon" consisted of several water barrels filled with G. *affinis* pulled by a team of horses. When inspectors found An. *quadrimaculatus* larval habitats, they seeded the problem area with minnows.[36]

Nearly twenty years earlier, John Smith conducted a series of experiments examining the effectiveness of G. *affinis* as a means of biological control. Smith noted that U.S. Fish Commission researcher had found the stomachs of dissected G. *affinis* filled with mosquito larvae. A year later, the New Jersey Fish and Game Commission blocked the request of Smith and Dr. Edward Rowe, Health Officer in Summit, New Jersey, to perform a series of experiments involving birds and top minnows. "Under the law," Smith wrote in his appeal, "I am expected to study the natural enemies of mosquitoes. To do this will require the killing of some birds that are either popularly credited with destroying mosquitoes or are known to feed on the insects."[37]

Howard Frothingham, president of New Jersey's Fish and Game Commission, denied both requests on the grounds that it would lead to the needless slaughter of birds and aquatic life forms. "While there is no doubt," Frothingham declared, "that top minnows will eat the larvae of mosquitoes, unfortunately the mosquitoes do not breed during most of the year, and the fish would have to live on something when there are no mosquitoes."[38] Despite this setback, Smith continued to explore the use of fish and other natural agents as instruments of mosquito control. "Of all the natural enemies of mosquito larvae," Smith noted in his 1904 *Mosquitoes of New Jersey*, "fish are the most important from the practical standpoint because they can be transported to places where they are needed, because they will stay where put and because they live throughout the season."[39]

Buck did not share the Frothingham's scruples about G. *affinis*. In rural Florida, top water minnows offered an inexpensive, effective means of eliminating anopheline larvae. In 1921 Raymond Turck congratulated Buck on his novel use of G. *affinis*. Turck, who had recently become Florida's state health officer, described how the tiny fish had rejuvenated Brewster. Buck's anti-mosquito campaign worked miracles. "I even sat out on an unscreened veranda step and wandered about in the bushes and on the edge of a pond at dusk trying to locate at least one mosquito. Not one. Not a bite."[40] Twelve months earlier Brewster had been a "pest hole." Turck maintained that Buck's anti-mosquito initiative had made Brewster one of the healthiest places to live in Florida.[41] Two years later Buck reported that the malaria rate in Brewster had plummeted to "a fraction of one percent."[42]

Simons hoped that the successful anti-malaria work in Perry and Brewster would lead to more support for mosquito control in Tallahassee. He was disappointed. In 1921 the governor slashed the state board of health's budget by 50 percent. Simons remained undaunted. If politicians failed to see the importance of the war against mosquitoes, Simons vowed to show them. He was convinced that the state's growth depended on winning the battle against mosquitoes.

An outbreak of dengue fever in 1921 provided Simons with an opportunity to prove his point. Epidemics of dengue fever periodically swept through the South in the nineteenth century. Dengue fever is typically a nonfatal, viral infection in which patients suffer from headaches, body pain, nausea, and vomiting. In 1903 Harris Graham, a member of the American University in Beirut's faculty, demonstrated that dengue fever was a mosquito-borne disease.

Three years later Thomas Bancroft, an Australian toxicologist notorious for his "self-satisfied infallibility, so characteristic of the young practitioner," used a group of volunteers to prove that *Aedes aegypti* mosquitoes could transmit the disease after a ten-day incubation period.[43]

In 1921 Miami suffered a severe dengue outbreak. Newspapers throughout the country carried reports of the epidemic. Physicians diagnosed more than five hundred cases. Hundreds more cases went unreported.[44] Fear spread through city. Businessmen worried that "break bone fever," as the illness was locally known, would deter tourists and end the real estate boom. In December, the city's commissioners asked Simons to prepare an anti-mosquito plan for Miami.

Simons presented his plan to the Miami City Commission in March 1922. Like Herms, Simons stressed the importance of winning the public's support for the mosquito work. An educated public was essential in waging war on mosquitoes. The commissioners agreed and allocated ten thousand dollars for a citywide anti-mosquito campaign. Simons spent much of the next four months rallying public support for the mosquito work.

In 1922 a dengue epidemic swept across the Gulf Coast. Physicians reported scattered cases in Florida in May. By August, dengue raged throughout the Southeast. Texas, Georgia, and Florida had the highest number of the 26,779 clinically documented patients. Health officials estimated that more than 200,000 people were infected.[45]

Simons later called the 1922 dengue epidemic a "blessing." Miami, Perry, and Brewster passed unscathed through the dengue outbreak. At the height of the epidemic, officials in Tampa and Jacksonville requested Simons's assistance. Both cities launched anti-mosquito campaigns. Confidence in public health was restored. The 1922 dengue epidemic demonstrated the effectiveness of anti-mosquito work. "It gave us an admirable chance," Simons explained to the members of the NJMEA in 1923, "to awaken interest in the whole program and start a statewide control program. The people were in a receptive mood to consider mosquito control measures, and today there isn't a community in Florida that wants to tolerate another dengue outbreak."[46]

Simons rallied support for the formation of a statewide anti-mosquito organization after the dengue epidemic. His message did not vary: Mosquitoes threatened the state's future. On December 6, 1922, 150 anti-mosquito warriors gathered in Daytona. "It is the plain and sober truth," Raymond Turck told the assembly, "that we humans are engaged in a battle for self-preservation." Turck challenged his listeners to join the mosquito crusade.[47]

The participants at the Daytona meeting answered Turck's call for action affirmatively. On December 7 they voted to organize the Florida Anti-Mosquito

Association (FAMA). They elected Joseph Porter to be the association's first president. Porter's election ensured the organization's credibility. His career was synonymous with the fight to eliminate mosquito-borne diseases. Raymond Turck introduced Porter's presidential address. "Twenty-five years hence," Turck observed, "we will look back to this momentous date in Florida's history. It will be remembered as the natal day of an organization which I am confidant, will bring about the realization of the titanic work of controlling, if not eradicating the mosquito from the state."[48] Taking the gavel, the seventy-five-year-old Porter echoed Turck's enthusiasm. The goal was clear. Victory over the mosquito was assured, Porter declared, if the people of Florida "Keep Everlastingly at It."[49]

New Jersey provided a model for the Florida mosquito fighters. "Our first step in the direction of this control problem," Simons explained three months later at the 1923 NJMEA annual meeting, "will be doubtless passage of certain legislation enabling committees to engage in control work outside the confines of their corporate area, possibly by providing for the creation of mosquito control districts of some character."[50] During the next three years, more than fifty towns organized mosquito leagues or implemented municipal ordinances mandating the enforcement of mosquito control measures.[51] In May 1925 the legislature authorized the formation of independent mosquito commissions. In June the newly formed Indian River County on Florida's east-central coast organized the state's first mosquito district.

The formation of the Florida Anti-Mosquito Association was part of a wider national drive for mosquito control in the early 1920s. In the midst of the 1922 dengue epidemic, LePrince laid the foundation for mosquito control in Utah and Illinois. Theodore B. Beatty, Utah's state health commissioner, asked LePrince to come to Utah. Concern about mosquitoes in Utah surfaced when the first settlers arrived in what is now Salt Lake City in July 1847. The early pioneers considered great hordes of mosquitoes to be acts of nature like droughts, floods, crickets, and grasshoppers. In 1922 Beatty and a handful of progressive citizens decided that the time had come to do something about the mosquitoes. Reports of the successful work in New Jersey and California inspired Beatty and a local attorney named E. W. Senior to invite LePrince to come to Salt Lake City.

LePrince surveyed Salt Lake City in 1922. Along with the survey, LePrince prepared a draft for a bill authorizing the formation of mosquito abatement districts on the New Jersey model. In February 1924 LePrince pointed to Utah as an example of the growing interest in anti-mosquito work. He described Beatty's novel efforts to stimulate the public's support for the mosquito crusade. "To arouse public interest in mosquito elimination," LePrince declared in an Atlantic City speech, "the state health officer of Utah writes his messages to people about mosquitoes with mosquitoes. The words are made with the mucilage and mosquitoes are poured on. All his messages are read."[52]

In 1923 Utah became the third state to pass legislation authorizing the formation of mosquito abatement commissions. The Salt Lake City Mosquito District formed in 1924. It held its first meeting in May and approved a salary of one hundred dollars per month for a "competent man" for the remainder of the mosquito season. Temporary work, oiling, continued in 1924. In 1925 the Salt Lake City Commission approved the district's first permanent ditching work along the Jordan Creek.

During the next decade, LePrince assisted the Utah movement. When duck hunters objected to the district's oiling, LePrince advised the commission explore biological means of control as part of a comprehensive mosquito control strategy. More than anything else, the Salt Lake City district needed a comprehensive entomological survey. LePrince told the commissioners that science must guide the district's anti-mosquito work. "All successful work done in the coastal counties of New Jersey," LePrince explained in 1928, "was based on a similar survey and, you know, more progress has consequently been made in that state in mosquito control than in all the other states combined."[53] LePrince suggested that the commission hire Professor Ralph Chamberlin, an entomologist at the University of Utah, to conduct the survey. "If this study is ignored," LePrince warned, "then sooner or later your committee may be severely and justly criticized for working blindly and wasting funds because of lack of knowledge of local entomological data."[54]

The Salt Lake City Commission accepted LePrince's counsel and hired Chamberlin to prepare the survey. LePrince dispatched one of his assistants, William Komp, who had been one of Headlee's first graduate students at Rutgers, to help Chamberlin plan the survey. Chamberlin, on his part, recruited a twenty-seven-year-old graduate student named Don Rees to serve as his aide. Chamberlin and Rees later discovered that the private gun clubs located west and northwest of the city that objected to mosquito control were the "principal source" of the salt marsh mosquitoes that plagued the city. "The swamps and marshes controlled by these clubs produce millions of mosquitoes each year," Chamberlin and Rees concluded.[55] It would take Rees many years to find a solution to this problem.

LePrince's efforts to promote the mosquito cause were not limited to Florida and the Mountain West. He dispatched a USPHS malaria investigation team to southern Illinois in 1922. French and British settlers had carried malaria to the Ohio and Mississippi river basins. At the beginning of the nineteenth century, malaria earned Illinois the reputation of being the "Graveyard of the Nation." In the early 1840s, Charles Dickens offered English readers a "lurid description" of his "journey by steamboat from Cincinnati to St. Louis." Dickens described southern Illinois "as a dismal swamp, on which the half-built houses rot away . . . a hotbed of disease, an ugly sepulcher, a grave uncheered by any gleam of promise: a place without one single quality, in earth on air on water, to commend it: such is dismal Cairo."[56]

After his American odyssey, Dickens presented a fictionalized version of Cairo, Illinois, to his readers in *Martin Chuzzlewit*. In the serialized novel, unscrupulous land developers convince Dickens's protagonist to purchase land in a fictional southern Illinois town called Eden. The real estate agents portray Eden as an earthly paradise. At one point, Chuzzlewit inquires if mosquitoes are a problem in Eden. He is told "there air [*sic*] some catawampous chawers in the small way, too, as graze upon a human being pretty strong; but don't mind *them*—they're company" [emphasis in the original]. Chuzzlewit and his friend discover that Eden is full of "noxious vapour" and "pestilential airs" that are "deadly poison." "Nobody," readers learn, "as goes to Eden ever comes back a-live."[57]

Malaria in Illinois was not limited to its southern counties. In the 1850s the disease raged in the new settlement of Chicago. Illinois health officials reported a malaria rate of roughly fifty deaths per hundred thousand in Chicago during this decade. The city's malaria rate declined in the 1870s and 1880s as land developers drained the swamps along Lake Michigan's North Shore.[58]

By 1920 "Little Egypt," as southern Illinois was known, stood in sharp contrast to Chicago. Illinois's twelve southern counties had malaria rates that bore a closer resemblance to Mississippi or Florida than to the upper Midwest. In 1922 the USPHS and the IHB supported a malaria demonstration project in Carbondale, Illinois. Located in Jackson County, Carbondale had a population of seven thousand in 1922. Seventeen years earlier Henry Mitchell, a Carbondale-based physician for the Illinois Central Railroad, called for the adoption of an anti-mosquito campaign to eliminate malaria. In 1916 and 1917 state health officers chronicled the disease's negative effect on the region's development.[59]

Early in 1922 Lyell Clarke, a young USPHS malaria worker, arrived in southern Illinois. Clarke, a civil engineer who had studied at Virginia Polytechnic Institute and Johns Hopkins University, began his career in mosquito control during World War I as a sanitary engineer in the USPHS extracantonment work. Later, he was part of an American relief team sent to Italy to aid local anti-mosquito workers. When he returned from Europe, LePrince assigned him to a malaria investigation unit based in Mississippi.[60]

Shortly after the New Year in 1922, LePrince made a hurried visit to Carbondale to organize the USPHS effort. During his stay, he collected a number of mosquitoes in the coatroom of the Hotel Roberts. He later sent the specimens to Leland Howard who in turn gave them to Harrison Dyar for identification. "The mosquitoes," Howard wrote to LePrince, "prove[d] to be all males of the yellow fever mosquito, which Doctor Dyar insists now be called *Aedes aegypti*. . . . Doctor Dyar thinks that these mosquitoes must have been breeding in the warm rooms of the hotel all winter. They must have come in on the person of some traveler from the south."[61]

After LePrince's visit to Carbondale, Clarke took charge of the USPHS anti-malaria work in southern Illinois. Clarke's immediate challenge was winning local support. Carbondale's Lions' Club pledged two thousand dollars. The IHB contributed one thousand dollars and the Illinois Central Railroad promised to correct drainage problems along the railroad's right of way. By year's end, mosquito workers had drained a sixty-acre swamp and dug more than four miles of ditches. The number of reported cases of malaria dropped from 267 in 1921 to 19 instances in 1922. A year later, there were only 11 cases. Carbondale's success inspired the Rotarians and Lions' Club in nearby Belleville to launch a similar initiative in 1923. Guided by a USPHS and IHB team, the Belleville anti-mosquito service clubs produced an 80 percent drop in the town's malaria rate.[62]

Newspaper accounts of the work in southern Illinois renewed interest in mosquito control along Chicago's posh Lake Shore Drive. When mosquitoes ruined a summer night concert in Highland Park, Spencer Fuller, a Riverside physician, launched a campaign for state support of mosquito control. In 1925 Edward Skinner, a USPHS mosquito investigator, surveyed suburban DuPage and Cooke Counties. Fuller used Skinner's survey to support his campaign for mosquito abatement legislation. In 1927 Illinois became the fifth state to adopt legislation allowing for the formation of mosquito abatement districts. Lyell Clarke left the USPHS to take charge of the newly formed Des Plaines Valley Mosquito Abatement District near Chicago.

Confidence in the movement's success rose to dizzying heights in the late 1920s. Armed with new control strategies, advocates of mosquito control proposed a sweeping expansion along the Gulf and Atlantic Coasts. This initiative grew out of a meeting in New Orleans in October 1925. During the previous summer, the entire Gulf Coast from Port Arthur, Texas, to Mobile, Alabama, was "invaded with enormous swarms of hungry mosquitoes."[63] New Orleans' mayor, Martin Behrman, invited other Gulf Coast mayors and state and federal officials to take part in a mosquito summit conference. Behrman, who led New Orleans during the 1905 yellow fever epidemic, had won reelection in May 1925 with the vow to rid the Crescent City of the mosquito pest.

In the weeks before the election, flights of *Aedes sollicitans* had descended on the French Quarter, disrupting weddings and driving revelers from Bourbon Street's speakeasies and jazz clubs. The salt marsh mosquito problem worsened during the summer. Louisiana's rogue mosquitoes left their Cajun home and paid unwanted visits to Mississippi and Alabama. Behrman's goal for the October meeting was to lay the foundation for a regional campaign against them.

Howard, who was beginning his fourth decade as the nation's chief entomologist, represented the secretary of agriculture at the New Orleans mosquito summit. He encouraged the participants to seek federal support. "No one state," Howard explained, "can be expected to attempt such a task as apparently confronts Louisiana. . . . The problem is distinctly an interstate problem, since Louisiana mosquitoes this last season caused more trouble in neighboring states than they did in Louisiana." After numerous speeches, the delegates "sensibly concluded that the first thing to be done is to have a mosquito survey made under the direction of competent experts to determine the exact character of the marsh breeding places and the predominant mosquitoes." The mayor of Biloxi volunteered to lead the effort to raise the anticipated $50,000 needed for the survey.[64]

In February Howard provided the delegates to the annual NJMEA meeting with an update on Gulf and Atlantic coast salt marsh initiative. When the effort to pay for the survey through private subscriptions proved illusory, Byron Harrison, senator from Mississippi, introduced a bill authorizing twenty-five thousand dollars for a joint USPHS and USDA survey. Congress approved Harrison's bill. The House and Senate allocated an additional ten thousand dollars for the survey in 1927.[65]

Thomas Griffitts led the two-year survey of the Atlantic and Gulf coasts. In his preliminary report, Griffitts described the extent of the problem. There were nearly six million acres of salt marsh from Virginia on the Atlantic Coast running to Texas on the Gulf. More than 60 percent of the marshes (3,381,500 acres) were located in Louisiana. Florida ranked a distant second with 680,000 acres followed by South Carolina (429,000) and North Carolina (427,000).[66]

The number of mosquitoes produced in Atlantic and Gulf Coast salt marshes was astounding. Along the Gulf Coast, Griffitts reported finding "larvae in concentrations of 2,650 per dipper." Moreover, the pest and nuisance mosquito problem was not limited to the summer months. Griffitts observed flights of mosquitoes "in every month of the year during 1927–1928."[67]

Given the magnitude of the salt marshes, many despaired of finding a practical solution. Howard and Griffitts disagreed. First, the Gulf and Atlantic states were not alone. They could draw on New Jersey's twenty-five years of experience in salt marsh work. "New Jersey," Howard declared, "had blazed the way and is leading the world in all practical aspects of mosquito control."[68] Second, Griffitts's preliminary reconnaissance had revealed that only a fraction of the salt marshes produced mosquitoes.

Two different kinds of salt marshes were present along the Gulf and Atlantic coasts. The vast majority were low marshes. Low marshes are defined as areas below the mean high-water tide line; high marshes are above this mean. Tidal action rendered the low marsh unsuitable for egg deposition. The high marshes represented the principal breeding sites of *Ae. sollicitans* and *Ae. taeniorhynchus* mosquitoes.[69] The *Ae. sollicitans* that had darkened the skies in 1925 originated in a relatively small number of acres of high marsh that were inundated with larger than normal spring and summer rains. Griffitts acknowledged that the problem was formidable but it was not "hopeless."[70]

Howard was even more confident. He had lost none of his zeal for the mosquito crusade. More states were joining the mosquito crusade. In 1928 Mississippi legislators approved an anti-mosquito law that gave counties along the Gulf Coast and Mississippi River authority to form abatement districts.[71] More importantly, Howard believed that recent developments in Mound, Louisiana, held great promise for the future of chemical control. In 1913 the Bureau of Entomology established an experiment station in Mound. Two years later, Howard recruited a newly minted Ph.D. named Willard van Orsdel King to work at the station. The twenty-seven-year-old King, who earned his doctorate at Tulane with a dissertation on "The Mosquitoes of New Orleans and Vicinity," was well on his way to becoming the foremost expert on the mosquitoes of the Southeast. At Mound, King led the research on new means of controlling salt marsh mosquitoes.

In 1920 Emile Roubaud, a French entomologist working at the Pasteur Institute, announced the discovery of a new larvicide. He reported that the chemical trioxymethylene killed anopheline larvae without producing harmful effects for fish. Roubaud's work encouraged Marshall Barber and Theodore Hayne at the USPHS to seek less expensive chemical means of control. In 1921 they revealed their discovery that Paris green, an arsenic-based poison long used in agriculture, was an effective larvicide.

Howard, whose career as a mosquito researcher began with his research on oil as a larvicide, was acutely interested in Barber and Hayne's discovery. In 1923 he reported the USPHS researchers' success using Paris green in a Lake City, Florida, field trial.[72] At the USDA Mound station, Willard King hit on the idea of using an airplane to apply the chemical on difficult-to-reach breeding areas. Marine pilots followed King's lead in 1926 at the Corps's Quantico, Viriginia, training facility. An aerial application of one pound of Paris green per acre transformed Quantico from being the "worst hole for malaria on the Potomac" into a healthful recruit station.[73]

The discovery of Paris green's effectiveness as a larvicide encouraged other research developments. The New Jersey Associated Executives in Mosquito Control (NJAEMC) formed in 1921. The associated executives' monthly meetings provided an opportunity for technical discussion. In 1925 Headlee recruited a young biochemist named Joseph Ginsburg to the agricultural experiment station. Ginsburg worked closely with the NJAEMC. Ginsburg and Robert Vannote, Morris County's superintendent, joint research effort culminated in the development of a pyrethrum-oil larvicide in 1933 that was known as the New Jersey Mosquito Larvicide.[74] Two years earlier, Vannote had followed King's lead and introduced aerial applications of pesticides in New Jersey.[75]

Howard's hope for finding a chemical solution to the Gulf and Atlantic salt marsh problem went beyond larvicides. Ellsworth Filby had contacted Howard in 1926 proposing a radically new form of mosquito control. Filby, who replaced George Simons as the sanitary engineer at the Florida State Board of Health, had written to Howard describing millions of mosquitoes that breed in the mangroves and coastal salt marshes. "It has occurred to him," Howard declared in a 1927 speech in New Jersey, "that some poison gas developed by the Chemical Warfare Service could be mixed with smoke clouds and blown over these areas destroying very many adults. The destruction of larvae in these areas he thinks impractical."[76]

Howard's subsequent investigations revealed that the Chemical Warfare Service had "developed a number of effective methods for the dispersion of poisonous smoke clouds." Howard wondered if poison gas might prove to be an inexpensive means of killing adult mosquitoes. In 1927 he invited the members of the NJMEA to consider "the practical use of something of this sort in mosquito warfare."[77]

Howard's invitation opened a debate that would prove to haunt the antimosquito movement. One member of the association wondered "if such a barrage [of poison gas] would be harmful for humans?" Headlee agreed that this was a legitimate concern. If a poison gas could be devised that was not toxic for humans, Headlee declared he had "no doubt whatever that . . . there are many places in New Jersey where it could be used." Others like William Donnelly from Flatbush, Brooklyn, worried that even if the gas had no effect on humans, it might prove fatal to birds and wildlife.[78]

Howard's support of research on poison gas as a potential tool in mosquito control because "all anti-mosquito work is [was] sanitary work"[79] drew criticism from Harrison Dyar. Dyar, who described himself "from a biological and taxonomical point of view . . . a mosquito lover," disputed Howard's claim. Howard reported at the 1927 NJMEA that Dyar considered the anti-mosquito crusaders "homo-centric." "You slash about," Dyar declared, "and don't care what becomes of the rest of creation so long as somebody can make a subdivision on a salt-marsh."[80]

Dyar's anger at the mosquito warriors "egotistic presumption" of the superiority of human beings over all other living things reached a boiling point in the 1920s. "There is nothing," Dyar confided to Howard, "I like less than killing a mosquito. I love to see the mosquitoes in vast swarms, and if I had my way, all the oil would be poured over the human exterminators." A visit to Yosemite and Yellowstone two years later left Dyar in a fury. "Every hollow that could not be drained or filled," Dyar wrote to Howard of his experience at Yosemite, "is full of stinking oil. Every marsh plant and hundreds of marsh insects are a thing of the past." The situation at Yellowstone was worse. "Why not do it right? Blast the whole place down with dynamite and steam shovels and plant it smoothly to lawn grass." For Dyar the addition of poison gas could hardly make things worse than they already were.[81]

These questions remained unanswered in 1926. Twelve years later Dyar's concerns about mosquito control's effect on the environment were at the center of a heated debate at the Third National Wildlife Conference. But by 1927 Dyar was a broken man. A decade earlier, rumors of bigamy led to his dismissal from his post at the USDA. A costly divorce settlement and his subsequent remarriage to his "spiritual advisor" contributed to Dyar's professional and social isolation.

Despite Dyar's biting criticism, Leland Howard remained Dyar's loyal friend. He believed that Dyar's skill as a taxonomist more than compensated for his intemperate outbursts. Howard persuaded the Carnegie Foundation to publish Dyar's six-hundred-plus page *Mosquitoes of the Americas* in 1928. The book contributed to restoring Dyar's professional standing. Dyar hoped that it would lead to his reappointment at the USDA. Howard, who had retired in 1927, agreed and supported Dyar's application. In January 1929 Howard's successor, Charles Marlatt, authorized Dyar's reemployment at the bureau. Dyar, however, suffered a stroke and died before his reappointment took effect. In February Howard eulogized his colleague and friend at the annual meeting of the New Jersey Association. "It will be years," Howard declared, "I fear many years, before he can be replaced by an American worker of even approximate knowledge and qualifications."[82] William Forbes put it even more succinctly in his obituary in *Entomological News*. "There is no one to take his place."[83]

Dyar's negative judgment of the anti-mosquito movement's contributions to "progress of civilization" did not diminish LePrince's and Griffitts's resolve.[84]

The mosquito warriors had accomplished much since 1912. By 1928 the USPHS had conducted mosquito investigations in twenty-four states in 667 counties.[85] In California, Herms and Gray had established an effective anti-malaria campaign. Mosquito workers were fighting pest mosquitoes in Illinois, Utah, and Florida. There was every reason to believe that the anti-mosquito movement would continue to grow. In April 1929 Griffitts reported on his salt marsh survey at the Gulf and South Atlantic Anti-Mosquito Congress. At the meetings' conclusion, the participants approved a resolution creating a regional anti-mosquito organization patterned after the New Jersey Association. Griffitts judged the new group a "wide awake organization."[86] The anti-mosquito movement was winning converts. Research on new chemical tools was promising and the future of mosquito control seemed bright. None of the mosquito crusaders anticipated the economic crash that began in October 1929. The next ten years brought challenges and opportunities that were destined to redefine the anti-mosquito movement.

Chapter 6 Advances and Retreats during the Great Depression

Six months before the October 1929 stock market crash, humorist and Oklahoma cowboy philosopher Will Rogers discovered the mosquito crusade. Thomas Headlee's presentation on "The Relation of Rainfall to the Seasonal 'Peak Load' of Mosquito Control" at the sixteenth annual meeting of the New Jersey Mosquito Extermination Association provided the unlikely occasion for Rogers's musings. Headlee had traced the breeding potential of a single female *Culex pipiens* mosquito through six generations. By the sixth generation, Rogers learned, a house mosquito could theoretically have produced nearly eighty billion descendants.[1] This astounding number prompted Rogers to quip "what we need . . . is pamphlets for the female mosquitoes on birth control. . . . Teach them that the days of the big families in mosquitoes are past. . . . Don't try to kill off the females. Educate 'em to modern ways."[2]

Despite Rogers's homespun whimsy, the anti-mosquito movement had accomplished much in the 1920s. Once an object of ridicule, mosquito crusaders were active in more than two dozen states. Six states (New Jersey, California, Utah, Florida, Illinois, and Mississippi) had passed legislation authorizing the formation of independent mosquito abatement commissions. Forty towns in Mississippi were conducting anti-malaria campaigns. In Texas the state department of health had appropriated ten thousand dollars to support anti-mosquito work in one hundred communities. Twenty-three counties in Alabama and forty cities and towns in Georgia had started control work. In Virginia Norfolk, Newport News, and twenty other towns had built on U.S. Public Health Service (USPHS) anti-malaria initiatives.[3] In New England in 1923 the Connecticut legislature amended the state's mosquito control law mandating that "the State pay *all* expenses in maintaining drained areas."[4] In neighboring Massachusetts twenty-one communities announced their intention to form the Massachusetts Mosquito Control Association in 1929.[5]

In New Jersey eleven counties had formed independent mosquito commissions. By 1926 the commissions were spending $325,000 annually to eradicate mosquitoes. The movement's leaders thought this figure would grow. Trenton lawmakers had recently amended the 1912 abatement law giving the freeholders (county commissioners) the right to issue bonds (up to $300,000) to support mosquito control work.[6] Veteran New Jersey mosquito fighters believed the bond money would give counties the resources to complete their salt marsh work years ahead of schedule through the purchase of a new generation of expensive, powerful ditching machines. In 1903 workers used shovels and ordinary garden spades in John Smith's drainage experiments in Monmouth County. During the next decade Harold Eaton, a mosquito inspector in Atlantic County, tested several different types of mechanical trenching machines. By 1914 Eaton had developed the Eaton Ditching Machine, a gasoline-powered engine mounted on two large metal drums that pulled a plow.[7]

Keeping the commissions' ditches in the tidal salt marshes open provided an incentive to develop a new generation of mechanical ditchers and ditch-cleaners. Under Headlee's leadership, researchers at the agricultural experiment station in conjunction with the New Jersey Associated Executives in Mosquito Control (NJAEMC) deployed a new generation of mechanical behemoths along the Jersey Shore. By the decade's end, massive ditching machines weighing twelve tons with a ten-thousand-dollar price tag and powered by five-ton Holt tractors had cut a latticework of ditches through the state's salt marshes.[8]

Frank Miller, Headlee's assistant at the experiment station, reviewed the mosquito brigade's accomplishments in the 1920s in an Atlantic City speech at the annual meeting of the NJMEA. "A more extended use of mechanical equipment," he explained, "the general return to backyard inspections, the use of better larvicidal oils, . . . and the spirit of cooperation and helpfulness that existed between the different county units, aid in large measure towards making last season 'the best ever.'"[9] Miller prophesied a bright future.

The Great Depression provoked tremendous changes within the anti-mosquito movement. The downturn forced drastic cutbacks. Prospects for mosquito control dimmed as the depression worsened. On Florida's east coast, Martin County dissolved its mosquito commission. In 1933 the Florida Anti-Mosquito Association cancelled future meetings "due to shortage in funds."[10] The ambitious plans for the Gulf and Atlantic Coast Anti-Mosquito Association were shelved.

Figure 6.1. Thomas Headlee with a New Jersey Light Trap in the 1930s. Photo courtesy of the Center for Vector Biology, Rutgers, The State University of New Jersey.

The movement reached its nadir in 1933. Four years of depression robbed communities of the resources and political will to support the mosquito crusade.

California and New Jersey present case studies of the depression's injury to the mosquito crusade. The minutes and discussion notes for the annual Conference of Superintendents and Trustees of Mosquito Abatement Districts in California document this. In December 1930 William Herms invited the leaders of the state's anti-mosquito movement to meet in Berkeley. Thirty individuals gathered in a classroom in Agricultural Hall in Berkeley for the California Mosquito Control Association's first formal meeting. Herms made certain that Henry Quayle was present. Quayle provided a tangible connection to the movement's origin. The Depression threatened to undo twenty-five years of hard work. Herms recommended that the participants form an association to exchange ideas, to discuss common problems, and to serve as an advocate for organized mosquito control. The participants agreed.[11]

Harold Gray's position as the CMCA's secretary allowed him to keep close track of the depression's effect on the abatement districts. In 1932 Gray warned of two distinct threats to the mosquito movement. First he noted a "determined drive everywhere for a reduction in taxes." Gray, who became the manager and engineer for the Alameda Mosquito Abatement District in 1930, had recently faced a particularly vexing challenge. At a hot summer meeting in 1932 the Alameda Board of Supervisors "yelled like hell about our poor little one cent tax." The supervisors asked the county attorney if they were required to levy the one-cent tax to fund the mosquito abatement district. The attorney advised that the supervisors "could levy in part only or not level at all and then let the District sue." The supervisors promptly slashed the abatement district's budget. If other counties followed Alameda's example, "it could result only in chaos," Gray warned, "and the ultimate breakdown of government."[12]

A second threat came from the state Tax-Payers Association. To reduce taxes, the Tax-Payers Association was lobbying for legislation that would consolidate each county's "numerous separate tax-levying bodies under . . . centralized control." Gray believed the proposal would lead to additional budget cuts. Worse, it would place "mosquito abatement work under the unrestrained political control of the Board of Supervisors."[13] This would result in "Supervisors [trying]. . . . to get us to employ their incompetent relatives and friends who need a job," Gray warned. "You know how it is. I've had plenty of experience with this sort of thing. . . . Politicians and their friends and relatives are seldom interested in slaying mosquitoes; their interest is almost solely in the salary warrant."[14]

The situation in New Jersey was, if anything, worse than in California. The depression threatened to undo three decades of work. The worsening economic conditions forced drastic cutbacks. By 1933 Atlantic County's freeholders had slashed its mosquito commission's budget by 93 percent.[15] Jacob Lipman, who had led the agricultural experiment station since 1911, summarized the situation. At the beginning of the depression, the extermination commissions spent $487,000 for mosquito control. This figure dropped to $243,500 in 1933. Hudson County experienced the smallest reduction. The four southernmost counties suffered the greatest losses. In Atlantic County, which advertised itself as the "World's Play Ground," the 1933 budget of $3,500 provided enough money to employ a part-time secretary and a superintendent.[16]

Lipman's numbers did not reveal human costs that could not be tallied in dollars. "Experienced mosquito fighters were not born," Fred Reiley, a sanitary engineer in Atlantic County, declared. "They have to be trained in the work before they can be of any value. In many counties . . . reductions have been so great, it has been necessary to drop experienced men."[17] It would take years to train replacements for these veteran mosquito workers.

In April 1933 Fred Bishopp, chief of the USDA's Division of Insects Affecting Man and Animals since 1926, offered a glimmer of hope to the struggling New Jersey mosquito warriors. "It seems to me," Bishopp declared in a speech in Atlantic City, "that the great public benefits to be derived from anti-mosquito operations and the excellent opportunity which it gives for utilizing large numbers of the unemployed cannot be too strongly emphasized."[18] Bishopp singled out Massachusetts as an example of how other states and the federal government could use mosquito control to provide jobs for unemployed. In Massachusetts, the state reclamation board's "first concerted action was on the island of Nantucket." In 1930 the legislature provided support to the Cape Cod Mosquito Project as part of the state's depression relief initiative.[19] Bishopp hoped that the new president would include mosquito control in his economic initiatives.

The New Deal redefined the relationship between the federal government and the American people. In his inaugural address, Franklin Roosevelt asked Congress for sweeping "power to wage a war against the emergency."[20] Mosquito control was a central component in FDR's effort to restore the nation. By December 1933 medical entomologists and sanitary engineers were planning a vast malaria control program for the Tennessee Valley Authority (TVA). Simultaneously, the Civil Works Administration (CWA) put thousands to work digging ditches and draining marshes. Federal subvention for local mosquito control continued until World War II.

The TVA was mosquito crusade's greatest achievement in the New Deal. On May 18, 1933, Roosevelt signed into law the bill authorizing the TVA. In his message to Congress proposing the legislation, the president declared his commitment to comprehensive, long-range planning. The two million people who lived in the six states in the Tennessee valley were among the nation's poorest. The TVA became a model of Roosevelt's hopes for the New Deal. The TVA, Roosevelt explained, "should be charged with the broadest duty of planning for the proper use, conservation, and development of the natural resources of the Tennessee River drainage basin and its adjoining territory for the general, social and economic welfare of the nation."[21]

The TVA had four major objectives: flood control, navigation, power development, and soil conservation. From its inception, the program's directors recognized that malaria posed a threat to TVA's success. Malaria was endemic in the Southeast. A plan that envisioned the construction of forty-eight dams with a string of impoundments covering six hundred thousand acres with eleven thousand miles of shoreline in six different states seemed a recipe for disaster in terms of mosquitoes and mosquito-borne diseases.[22]

Speculation over the relationship between such impoundments and spikes in the malaria rate first surfaced in 1910 in central Alabama after the completion of a small water reservoir on the Cahaba River. Two years later, Henry Rose Carter and what became known as the USPHS "old malaria gang"[23] investigated suspicious outbreaks of malaria near a newly created pond in Blewett's Falls, North Carolina, and at hydroelectric impoundments on the Tennessee River and the Coosa and Warrior rivers in Alabama. Completion of the Hales Bar dam on the Tennessee River and the subsequent flooding of the reservoir basin without clearing the impoundment area provoked a public outcry over the increased number of mosquitoes. Residents filed a handful of unsuccessful lawsuits against the power company, claiming the company was responsible for the increase in malaria that followed the dam's completion.[24]

The outcry against the power companies peaked during World War I. In 1914, a hydroelectric company's impoundment on Coosa River led to a malaria epidemic in Coosa when the reservoir was flooded. Encouraged by local attorneys, a number of individuals sued the power company alleging that the company's negligence had caused the malaria epidemic. The Coosa lawsuits were noteworthy because a local judge ruled in favor of residents' complaints. This decision was later reversed when the Supreme Court of Tennessee ruled that there was insufficient evidence to support the link between impoundments and malaria.[25]

Between 1915 and 1925 Carter and Griffitts documented the connection between impoundments and malaria. The USPHS proposed a set of regulations governing impounded waters in 1915. In 1921 Griffitts drafted the Federal Power Commission's guidelines for hydroelectric impoundments. The next year Alabama's legislature adopted Griffitts's recommendations penalizing individuals who created larval habitats for mosquitoes. Virginia and other southern states approved similar legislation.[26]

Thus, when Roosevelt signed the bill authorizing the TVA, there was widespread awareness of the malarial threat posed by impounded waters. Malaria was endemic in the region. Unless corrective measures were put in place, there would be a widespread outbreak of the disease in the region. Fortunately, there was also a substantial body of research demonstrating that certain water management strategies could significantly reduce the malaria's menace. Carter's and Griffitts's research demonstrated that mosquito control could significantly reduce the risk of malaria outbreaks.

Edward Bishop, Tennessee's former health officer, took charge of TVA's Health and Sanitation Section. By November 1933 Bishop had prepared "plans for a comprehensive malaria control program."[27] Drawing on Carter's, Griffitts's, and LePrince's studies, Bishop formulated a "naturalistic control" strategy for the TVA. Bishop recognized that there was no single remedy for malaria. Permanent shoreline improvements, fluctuations (both seasonal and periodic) of the water level in the impoundment, use of G. *affinis*, and limited application of larvicides offered means to eliminate anopheline breeding. Screening houses supplied an additional line of defense. Finally, TVA health workers distributed quinine to households near the reservoirs. Each impoundment was unique. The key was to remain flexible and employ a mosquito control plan that was tailored to each site's ecology.

TVA entomologists and engineers began anti-mosquito in 1934 at the Wilson Dam. Bishop's inspectors performed a malaria survey in the communities near the Dam. About 8 percent of the 1,693 blood smears drawn were positive.[28] Bishop ordered that the shoreline of be cleared of vegetation, robbing *An. quadrimaculatus* habitats during the spring. In the summer Bishop's inspectors applied Paris green on the Wilson reservoir to reduce the anopheline population.

The TVA's great strength lay in careful planning, research, and the cooperation between the different parts of the project. To guide the anti-malaria work, Bishop assembled a five-person review board: Louis Richards Jr., Joseph LePrince, and Thomas Griffitts from the USPHS; Willard King from the USDA; and Mark Boyd from the Rockefeller Foundation. "The value of the advice and assistance of these men," Walter Stromquist, the TVA's principal sanitary engineer, declared, "based upon their wide experience, cannot be overestimated."[29]

In 1933 the TVA offered little to address the nation's immediate economic agony. It would take years to revitalize the South. With millions unemployed and tens of thousands living in makeshift Hoovervilles, the president was eager to generate jobs. The administration's challenge was finding work for the millions of unemployed. During the spring and summer, the Federal Emergency Relief Administration (FERA) and its adjuncts became the locus of Roosevelt's jobs and relief effort.

The Public Works Administration (PWA) was one of FERA's first efforts to help states and local governments by providing loans for construction projects. Even the most modest PWA construction project, however, entailed lengthy delays for planning, acquiring bids for construction materials, and securing necessary permits. The situation reached a crisis point in the fall. Unwilling to allow the bleak economic situation to worsen, Roosevelt ordered the immediate transfer of four hundred million dollars from the PWA to the Civil Works Administration (CWA). The CWA represented a radical shift in federal policy. For the first time, the federal government assumed responsibility for putting the unemployed to work.

Mosquito control was the great beneficiary of the CWA and its successors. Six months earlier Fred Bishopp had hoped for a New Deal for mosquito control. In April he could not have anticipated the magnitude of the federal initiative. Roosevelt put Harry Hopkins in charge of the CWA. Hopkins asked the USDA, the USPHS, and a handful of other federal agencies for a list of projects that could provide immediate employment. Bishopp argued that mosquito control could provide jobs for hundreds of thousands of workers. Hopkins agreed and assigned Bishopp and Louis Williams Jr. to supervise the CWA mosquito-control projects.

In November Fred Bishopp briefed Headlee and the NJAEMC on the CWA program. Bishopp explained that there were, in fact, two CWA anti-mosquito initiatives. Williams, a seasoned anti-malaria physician at the USPHS, headed the CWA's malaria control program in fourteen southern states. Bishopp, who had joined the USDA in 1904 and earned his Ph.D. in entomology at Ohio State in 1932, directed the larger CWA pest mosquito control program in thirty-three states stretching from Maine to California. The goal, Hopkins stressed, was to put people to work.

The USPHS malaria drainage project in the South was one of four CWA health initiatives. The other projects included constructing sanitary privies, conducting typhus surveys, and sealing abandoned mines. Hopkins gave Williams $4.5 million dollars to launch the program. Ditching operations began in early December, and by January 20, 1934, 28,000 men were at work. Supplemental FERA grants to southern states allowed an additional 53,000 men to join the ditching effort. At its high point, USPHS officials estimated that 120,000 workers were digging ditches in the South.[30]

The swiftness and enormity of the work was astonishing. In fewer than one hundred days, CWA recruits dug more than six thousand miles of ditches. One hundred thousand acres of ponds and two hundred thousand acres of swamps were drained by February 1934. It was impossible, Williams admitted, to gauge the work's environmental consequences. Williams maintained that public health outweighed the harm to wildlife habitats. He claimed that the CWA anti-malaria work had significantly reduced the "hazard of malaria" for the "bulk of the 10,000,000 people adjacent to these projects."[31]

Williams overstated the health benefit. Incompetence was widespread in the USPHS CWA malaria drainage projects. Poor planning and the shortage of experienced personnel reduced the work's effectiveness. "On such short notice," Williams conceded, "it was not possible to organize technical supervisory forces in fourteen states."[32] This meant that USPHS state directors had to rely on inexperienced personnel to plan and manage the projects. Local politicians cared more about paychecks than environmental damage.

Louva Lenert, who prepared the Rockefeller Foundation's review of California's mosquito commissions a decade earlier, warned that without regular maintenance, ditching would not lower the malaria rate. "Malaria control through the [CWA] drainage program alone is not expected to have great permanent value, however, unless it is followed by continued maintenance."[33]

The CWA pest mosquito project presented a greater challenge. Bishopp's mandate stretched from New York to California. Most states lacked any experience with mosquito control. Bishopp, however, remained optimistic. He told Headlee and the leaders of the New Jersey he was convinced that "mosquito control work is one activity which has been promoted rather than handicapped by the depression."[34]

Finding experienced mosquito warriors to lead the army of ditch diggers proved a difficult task. By early December Bishopp had recruited a cadre of veteran control workers to supervise the twenty-two thousand men who dug more than 1,864 miles of ditches in the CWA pest work. Lester Smith, Roland Dorer, and Ralph Vanderwerker from New Jersey; Norman Platts from Florida; and H. Myhre from Massachusetts guided salt marsh ditching along the Atlantic Coast. Bishopp's colleagues in the Bureau of Entomology, Willard King, W. E. Dove, George Bradley, and Harry Stage, coordinated CWA efforts in Florida, Alabama, Mississippi, Texas, and Oregon. In California, Bishopp asked William Herms to oversee the CWA pest work "in 29 cities and 20 inland and coastal counties."[35]

These veteran mosquito warriors were an exception. More often than not, individuals with no experience led the CWA pest mosquito projects. "Hardly any of the states included in the federal [pest mosquito] projects had ever attempted mosquito work before," Lester Smith, the superintendent for the Middlesex Commission, observed, "and very few of them had any surveys made in the past from which they could receive guidance." Numerous projects lacked the guidance of trained, competent men. Worker efficiency remained low.[36]

On February 15, 1934, Hopkins ordered an abrupt halt of all CWA programs. Historian William Leuchtenburg argues that Roosevelt feared that the CWA was "creating a permanent class of reliefers [*sic*] whom he might never get off the government payroll."[37] Federal support for mosquito control continued. FERA, the CCC, and the Works Progress Administration (WPA) funded mosquito projects until World War II.

The New Deal's support for mosquito control was a mixed blessing. Federal dollars sustained established mosquito control programs in New Jersey, Florida, California, Illinois, and Utah. New programs were launched in two dozen states. The money came with stipulations. No federal dollars could be spent on maintenance of existing mosquito projects. Washington mandated that federal funds could be used only on new ditching and drainage projects.

Bishopp shared Hopkins's and Roosevelt's fear that individuals and local governments would become dependent on Washington. By 1934 Bishopp had noted a disturbing trend in New Jersey. The influx of federal dollars served as an excuse for mosquito abatement commissions to cut their budgets. This produced a serious problem. Municipal and county agencies cut funding needed to maintain completed drainage projects. "Failure to keep the ditching systems clean and working," Bishopp warned, "may result in extensive mosquito breeding and thus nullify the work." Successful mosquito control required "a well trained and effective organization. . . . Emergency funds, no matter how generously applied, will not do away with the necessity of continuous and consistent support."[38]

Utah, Delaware, and Suffolk County, New York, illustrate the positive and negative effects of the federal mosquito initiative. In Utah, Bishopp asked George Knowlton, an entomology professor at Utah Agricultural College, to organize the CWA effort. Knowlton received a $603,273 allotment to hire one thousand men for mosquito work. Six counties participated in the program. Knowlton directed the majority of the federal funds, $416,990, to the Salt Lake Mosquito Abatement District (SLMAD) where a young thirty-two-year-old named Don Rees served as the SLMAD's part-time supervisor.[39]

Don Rees was the catalyst for mosquito abatement in Utah.[40] Born in 1901, Rees earned his bachelor of science (1926) and master's (1929) degrees in zoology from the University of Utah. Rees's interest in mosquitoes grew out of his graduate studies. In 1928 Joseph LePrince recommended that SLMAD conduct a scientific survey. The district's trustees asked Ralph Chamberlin to lead the study. Chamberlin assigned Rees the job of doing the fieldwork. Rees based his master's thesis, "An Investigation on the Mosquitoes of Salt Lake County," on this work. After completing his degree, Rees took charge of the SLMAD as the district's part-time supervisor (1930–1937).

Rees guided the CWA's mosquito work in Utah. Rees's scientific background and practical experience made him an invaluable resource. In the midst of supervising the Salt Lake CWA work and guiding the six other CWA county projects, Rees continued to collect data on the state's mosquitoes. William Behle, who was later Rees's colleague at the University of Utah, recalled that when the CWA work ended, Rees "started scouting around for a suitable institution where he could quickly obtain a Ph.D."[41]

A quick trip to California, where he met with William Herms at Berkeley and visited Stanford, led Rees to enroll in Stanford for his doctoral studies. He completed his doctorate degree in one year (1935–1936) with a dissertation on "The Mosquitoes of Utah." Rees published an expanded version of his dissertation in 1943. In 1937 Rees resigned as the SLMAD supervisor and accepted a full-time teaching position at the University of Utah. As a professor, Rees continued to guide mosquito control's development in Utah while training more than one hundred graduate students in entomology. Many of Rees's students—Glen Collett and Lewis Nielsen, to name two—went on to distinguished careers in entomology and mosquito control.

The Utah CWA project's success grew out Rees's careful fieldwork. By 1933 Rees had formulated a detailed drainage plan for the Jordan Creek and the salt marshes east of the Wasatch Front. Without federal funds, Rees estimated that it might have taken as much as twenty years to finish the work. "The launching of the Federal plans for employment of large numbers of men," Ralph Chamberlin explained, "made possible the essential completion of this program and its extension in some respects."[42]

Federal pest mosquito work in Delaware provides a contrast with the Utah experience. In Delaware, interest in mosquito control was a recent thing. Nothing had come from John Smith's 1905 effort to encourage the Delaware governor and the state university chancellor to support a statewide mosquito program. In 1931 interest in mosquito control reawakened. A year later entomologists at the state agricultural experiment station prepared a survey of the state's mosquito population.[43]

Bishopp encouraged the experiment station to apply for federal assistance in the summer of 1933. He had learned that Robert Fechner, head of Federal Conservation Work, might be willing to assign Civilian Conservation Corps (CCC) recruits to mosquito control work. On September 21, 1933, Bishopp and Williams presented their case to Fechner for why the CCC should include pest mosquito control "in lieu of forestry work." Bishopp argued that "eradication work was much needed" and offered "an ideal hand labor project." He assured Fechner that the experiment station's entomologists had done the necessary preliminary work. New Jersey would provide a model for Delaware's mosquito effort. All that was needed was Fechner's approval.[44]

In early October Fechner assigned two companies of CCC workers to Delaware for mosquito work. Fechner stipulated that the state government must demonstrate its commitment to mosquito control. It took six weeks to convince the Delaware State Assembly "to appropriate the necessary funds for the executive headquarters of the mosquito control work." The debate over a $14,400 appropriation exposed "prejudices, skepticism, and patronage demands." Delaware's governor resolved the dispute by creating a five-member, nonpartisan Mosquito Control Commission.[45]

Delaware's pest mosquito work got off to a slow start. Before the 430 CCC recruits were fully employed, Bishopp sent a telegram to William Corkran, the Mosquito Control Commission's executive officer, announcing his decision to assign an additional 555 CWA workers to Delaware. Finding boots, tools, and other supplies for a thousand men proved, Corkran observed, "quite a little problem."[46] CCC mosquito control work continued in Delaware after the CWA ended in February. By November 1938 two companies of CCC workers dug 2,199 miles of ditches draining nearly half of the state's marshes.[47]

The Delaware CCC mosquito control experience exposed the friction that often developed between local and federal authorities. The speed in which the program took shape and the need to make decisions about deploying workers ensured that there would be bruised feelings. Local officials resented Corkran's way of doing business. Louis Stearns, entomologist at the Delaware Agricultural Experiment Station, praised Corkran for his ability to "marshal and keep engaged the large forces at his command." He failed, however, to build local support for the program. Stearns admitted that "the fact remains that the procedure which he was permitted to follow and the plans for the future of this effort were obviously not in line with the rural ways and thinking in Delaware."[48]

Management troubles were not limited to Delaware. In New Jersey entomological and engineering disputes between CCC officials occurred frequently. After the CWA's termination, Thomas Headlee persuaded Robert Fechner to assign eight CCC companies to New Jersey for mosquito control.[49] Suitable campsites were found, but shortages of lumber and building materials delayed the opening of the first CCC camp until July 1935. Many CCC camp superintendents, Fred Reiley declared, "were very frank in telling the mosquito commissions that they, the superintendents, would lay out the work, and that when it was completed the [mosquito] commissions would have to accept it."[50]

Additional problems grew out of the ballyhoo that accompanied the mosquito work. In Delaware Stearns noted from the "very outset of the utilization of CCC manpower" a pattern of exaggeration and overstatement. Stearns told his colleagues in New Jersey that the CCC "achievements, even minor ones, were over publicized."[51]

Press reports trumpeting the CCC's mosquito work encouraged Delaware's legislature to slash funding for mosquito control. In Wilmington, state officials came to consider the federal mosquito work as an entitlement and discontinued funding for maintenance work. In May 1939 Robert Fechner "ordered the discontinuance" of CCC activities in Delaware "as a disciplinary action following the failure of the legislature to provide sufficient funds for the maintenance of mosquito-control work." In August the Assembly passed a bill dissolving the state Mosquito Control Commission, transferring all "machinery, tools, and equipment" to the state highway department. This decision, state entomologist Stearns observed, "ended a most interesting chapter in the history of mosquito-control work in Delaware. As to the future of the movement," he told the delegates at the 1940 NJMEA annual meeting, "your guess is as good as mine."[52]

Federal support for mosquito control in Suffolk County, New York, was a disaster. Thirty years earlier Henry Clay Weeks pioneered the first mosquito control work on Long Island on Sheepshead Bay. Weeks led a crew of twenty men in "draining and petrolizing" what he termed "one of the finest mosquito breeding grounds in the United States."[53]

In 1916 the Nassau-Suffolk Mosquito Control Association won passage of a bill authorizing the formation of a mosquito abatement commission in Nassau County.[54] Mosquito control in neighboring Suffolk County languished until 1933 when local officials organized a mosquito abatement commission. Spurred by the idea of receiving federal support for mosquito control, Suffolk County officials used their political connections in Albany and Washington, D.C., to apply for FERA and CWA funds. Between June 1, 1933, and February 15, 1934, Suffolk County received $547,000 from the federal government for pest mosquito control. During this period, relief workers dug a latticework of nearly eight million feet of ditches through the county's salt marshes and wildlife preserves. No thought was given to the environmental consequences of the undertaking.[55]

Arthur Froeb, the Suffolk Commission president, admitted his own and his fellow commissioners' inexperience and ignorance. "I am not a scientist," Froeb declared, "and when I first commenced to fight mosquitoes I used to listen to these names, *Culex pipiens, Anopheles crucians, Aedes sollicitans*, and what not, and wonder what it meant in English." In 1933 Froeb and his fellow commissioners "knew very little about fighting mosquitoes but we did know that the 'doggoned' things sting like the devil. We had a job to do and set out to do it."[56]

No one on either the Suffolk or Nassau counties' mosquito commissions took notice in 1933 of a young Long Island wildlife reserve manager named William Vogt. Born in 1902 in Mineola, Long Island, Vogt was to become the mosquito crusaders adversary in the 1930s. Long before it became fashionable, Vogt described himself an ecologist. "My field properly speaking,"

Vogt observed in a 1945 article in the *Saturday Evening Post*, "was ecology and behavior of birds, but bird behavior and ecology are first cousins of human behavior and ecology."[57]

Visits to rookeries and nesting sites on the nearby coast fueled Vogt's passion for wildlife. After graduating from St. Stephens (now Bard) College, Vogt flirted with the idea of becoming a poet and a novelist. He found his life's work in 1932 when he became the curator of the Jones Beach State Bird Sanctuary on Long Island and editor of *Bird-Lore* magazine. Two years later, Vogt noted a 40 percent drop in the sanctuary's black duck population. Although alarming, Vogt cautioned that it was "too early" to reach any conclusions.[58]

Vogt left Jones Beach in 1935 to serve as the Audubon Society's national spokesman on behalf of the dwindling bird population. His frequent speaking engagements along the East Coast allowed him an opportunity to observe firsthand the growing wildlife crisis. An unvarying pattern had emerged from his trips along the East Coast. Wherever mosquito control workers drained salt marshes, bird and wildlife populations plummeted. In 1937 the Audubon Society published Vogt's thirty-two-page indictment of the mosquito crusade. "When we turn to a consideration of mosquito control," Vogt railed, "we encounter what seems perilously close to destructive, government-sponsored rackets. Much of the work is directed by engineers who have little knowledge of the biological problems and values involved."[59] Mosquito control, Vogt maintained, was "wildlife enemy Number 1."[60]

In Washington, D.C., a junior biologist in the United States Biological Survey (USBS) named Clarence Cottam had also noted the alarming drop the duck population. Born in 1899 in St. George, Utah, Clarence Cottam spent his first thirty years in Utah where he earned his bachelor of science (1926) and master's (1927) degrees from Brigham Young University. In 1929 Cottam joined the USBS (later United States Fish and Wildlife Service). Cottam's ornithological skill and passion for wildlife impressed J. N. "Ding" Darling, the Survey's chief. "The two men," Cottam's friend Eric Bolen remembered, "soon formed an admiration for each other that was to last far beyond their government service."[61] In 1934 Darling named Cottam the senior biologist in charge of the Section on Food Habits at the Survey (1934). Two years later, Cottam received his doctorate from George Washington University for his study of the "Food Habits of North American Diving Ducks." Later he was asked if he had personally examined the stomach contents of the seven thousand ducks in his study. Cottam, a Morman abstained from coffee, alcohol, and swearing, declared "Every D-A-M [spelling the word] one!"[62]

Cottam's painstaking attention to detail made him a formidable adversary; his dry wit made him lethal. In the midst of one of his recurrent battles with the Army Corps of Engineers, Cottam declared that the Corps working motto was "If it's wet, drain it; if it's dry, dam it." Cottam saved his sharpest barbs for Bishopp and the Bureau of Entomology. The problem with the Bureau of Entomology, Cottam maintained was their habit of "scalping the patient to cure the dandruff."[63]

Criticism of mosquito control was not limited to a sharp-tongued birder and a droll zoologist. Sportsmen joined the controversy in 1935. Hunters and fishermen meeting in New York between January 21 and 23 charged that the federally funded mosquito control projects were responsible for the dwindling numbers of waterfowl on the Atlantic seaboard. Matters became more serious when opponents of Suffolk and Nassau counties' mosquito control work sought to introduce legislation that "was adverse to mosquito control" during the 1935 Assembly session.[64] Their effort failed, but Bishopp and Headlee feared their opponents might someday succeed.

Six weeks later the beleaguered leaders of the New Jersey Mosquito Extermination Association (NJMEA) assembled in Atlantic City for their annual meeting. Charles Forbes, the association's president, rallied the participants in his presidential address. The mosquito crusade that had begun in the New Jersey marshes had spread across the continent. The NJMEA's influence reached far beyond the Garden State. "This convention," Forbes declared, "centralizes mosquito work on the North American continent."[65] Federal emergency relief funds had allowed an unprecedented expansion of mosquito control. These gains, Forbes admitted, had generated criticism. Forbes hoped that this meeting would help resolve these controversies. "I have come to the conclusion," Forbes declared, "that we have got to be more than usually patient and more than usually clear in our explanations."[66]

In the weeks before the meeting, the NJMEA's leaders had mapped out a strategy to dampen the growing controversy between organized mosquito control and the wildlife advocates. The NJMEA's resolution committee prepared a multiparagraph resolution that called for a "comprehensive scientific study" of mosquito control's relationship to the "continent-wide reduction in numbers of wild ducks and other migratory waterfowl, of shore birds, and of many other forms of wildlife."[67] The NJMEA resolution called for the creation of a committee drawn from the USDA Bureau of Entomology and the U.S. Biological Survey. "The situation," the NJMEA resolution declared, "constitutes a national emergency that is of immediate and personal concern to every mosquito fighter, every sportsman, every naturalist, and every lover of wildlife in America." Cooperation was the key to success. The NJMEA convention approved the resolution declaring their willingness to work with wildlife advocates in the future.[68]

The controversy over mosquito control intensified in the spring and summer. In June Headlee, Robert Glasgow, and Louis Stearns, the respective state entomologists for New Jersey, New York, and Delaware, sent out invitations to mosquito control workers on the Atlantic coast for an emergency meeting. Their goal was to formulate a common defense for the mosquito crusade against the movement's environmental foes.

Twenty-six mosquito workers from Rhode Island (one), Connecticut (one), Delaware (three), New York (six), and New Jersey (fifteen) assembled in Trenton on June 26. The attendees asked Headlee to preside over the two-day meeting. Tommy Mulhern served as the meeting's recording secretary. At the meeting's conclusion, the participants voted to form a common front to combat the spurious allegations against mosquito control. A new organization, the Eastern Association of Mosquito Control Workers (EAMCW), would lead the counterattack against the "individuals and organizations" spreading the falsehood that "that mosquito control operations have been an important factor in the depletion of wildlife."[69]

The Trenton meeting exposed Headlee and Bishopp's fear that the U.S. Biological Survey might obstruct mosquito control work in the future. Both the Biological Survey and the Bureau of Entomology were housed in the USDA. Wildlife activists had seized on the NJMEA's March resolution calling for an "inter-bureau committee of the United States Department of Agriculture." They saw it as an opportunity to wrest control of the federal mosquito work from Bishopp and the Bureau of Entomology. In Trenton, the newly formed EAMCW rejected the idea that Cottam or any member of the Biological Survey would have veto power within the interbureau committee. "Mosquito control organizations," the EAMCW members retorted, "cannot accept any plan that would result in divided authority." The interbureau committee was a consultative entity. Final authority for federal expenditures for pest mosquito control work must remain under Bishopp's direction at the Bureau of Entomology.[70]

During the next eighteen months, the wildlife controversy intensified. Delaware's salt marshes were the focal point for the dispute. The editors of *Delaware Sportsman* devoted extensive coverage to the issue in 1936. In a pair of articles, John Cahalan and William Vogt outlined the case against the mosquito control. A visit to Delaware had confirmed Vogt's fears. A field study had revealed "tragic evidence of the destructiveness of mosquito control." Everywhere Vogt went he found plants dying and "the destruction of a superb wild-life feed and shelter area by mosquito ditches."[71] Sentiment against mosquito control was growing. In Delaware Vogt claimed that support for mosquito control had dwindled. "The rumors," William M. Foord, president of the

Henlopen Game Farms Inc. in Milton, Delaware, declared, "you [Vogt] have heard about mosquito control are correct, and [I] believe there has been a great change here and in other parts of the state regarding it. . . . We took a poll and found that about 75% of them were now against ditching and would not permit it again if they had known what they know now."[72]

A year later, the Audubon Society published Vogt's thirty-two-page *Thirst on the Land*. J. N. "Ding" Darling, who had stepped down as chief of the USBS, provided four cartoons to illustrate Vogt's jeremiad against the nation's water management practices. "It remains to be seen," Darling had declared before leaving the USBS, "whether we have intelligence and ingenuity enough to repair the damage we have done."[73]

Federal support of mosquito control ditching was but the most recent assault on the nation's water resources. "The zeal for drainage," Vogt maintained, "has ravaged the North American continent like some form of terrestrial erysipelas. Wildlife and vegetation have been killed, and the earth has dried up."[74] Vogt argued that the federal mosquito control work was a "publicly financed sales talk . . . that under the guise of 'education' officials concerned with mosquito control have made sweeping claims without producing any scientific proof." Citing the work of Clarence Cottam, Warren Bourne, and F. M. Uhler at the Biological Survey, Vogt charged that the CWA-, CCC-, and WPA-sponsored mosquito control was no more than a "government-sponsored racket."[75]

The dispute reached a boiling point when the USBS blocked a pest mosquito control project in Rhode Island. Milton H. Price, supervisor of Rhode Island's pest mosquito work, reported that early in 1937 all of the projects he had submitted to the WPA were returned "with the notation, 'disapproved by the United States Bureau of Biological Survey.'" The work remained stalled until Cottam and Bertrand Smith visited Rhode Island and reviewed the proposal. Cottam and Smith "approved the projects, with reservations, and within two weeks," Price told an Atlantic City audience of mosquito fighters, "we were able to begin work on projects that had previously been disapproved."[76] For the first time, wildlife advocates succeeded in blocking and modifying a mosquito control project.

The CWA, CCC, and WPA mosquito projects created a legacy of hostility between advocates of mosquito control and an increasingly vocal group of hunters, fishermen, and wildlife advocates. The Third Annual National Wildlife Conference held in Baltimore in March 1938 revealed the depth of the distrust between the mosquito warriors and the wildlife advocates. J. N. "Ding" Darling had organized the first National Wildlife Conference in 1936. The Third Conference in 1938 devoted a special session to a debate over the

Figure 6.2. J. N. "Ding" Darling cartoon from William Vogt's *Thirst on the Land*, "The U.S. Public Health Service Takes a Shot [at Mosquitoes]." Reproduced courtesy of the Ding Darling Wildlife Society.

question, What Is Wrong with Mosquito Control. Clarence Cottam and William Vogt presented the case against mosquito control. Fred Bishopp and Louis Williams Jr. delivered the mosquito crusaders' rebuttal.

Cottam spoke first. He charged that mosquito drainage and ditching projects showed a pattern of "*improper* practices in attempts to exterminate mosquitoes" [emphasis in original]. Since 1933 federal malaria and pest mosquito projects had drained one million acres of wetlands. Pest mosquito salt marsh drainage accounted for half of this total. This drainage had led to the wholesale destruction of nesting and feeding habitats for birds, fish, and crustaceans. "A large percentage," Cottam declared, "of mosquito-control projects as hitherto conducted in this country has been unnecessarily destructive to wildlife."[77]

The use of oil as a larvicide posed the second great threat to wildlife. Cottam declared that the use of oil "cannot be too seriously condemned by all those interested in wildlife conservation." Cottam's research had revealed that the standard mosquito control practice was to use twenty gallons of oil per acre of salt marsh. Undoubtedly, this was a boon to cash-strapped petroleum companies in the midst of the Depression. It was, however, a disaster for waterfowl, muskrats, and oysters. In 1936 federal emergency funds had been used to purchase two hundred thousand gallons of oil for use on salt marshes in the northeast. Almost 25 percent of the oil was used in Suffolk County. There was, Cottam declared, no excuse for such extensive use of oil when potentially less harmful alternatives were available.[78]

Fred Bishopp and Louis Williams discounted Cottam's arguments. "No intelligent person can logically contend," Bishopp declared, that ". . . mosquito control as such is undesirable."[79] If there was a valid charge that could be made against mosquito control, it was that "there isn't enough of it, in the first place; in the second place, there are not enough people doing it."[80] Neither man addressed Cottam's specific charges about the decrease in plant life and the drop in wildlife populations that accompanied drainage and oiling. "A few ditches," Williams conceded, might have been dug in the wrong places. That number, however, was "so small as to be negligible." Cottam and Vogt had ignored mosquito control's tangible benefits. Property values had increased and the malaria rate had declined. "If the anti-malaria drainage work is wrong," Williams concluded, "it will come to a close; if those who object to it are wrong, anti-malaria drainage will continue, because things that are right have a way of prevailing in the long run."[81]

Bishopp's and Williams's defense of mosquito control gave Vogt an opportunity to display his talent as a debater and public speaker. He had used the weeks prior to the debate to review the previous twenty-five years of the *Proceedings of the New Jersey Mosquito Extermination Association*. Vogt's tone was clear from his opening. "The Audubon Society, like the Biological Survey, holds no brief for mosquitoes. We admit at the outset that nothing can be found wrong with *some* mosquito control" [emphasis in original]. The problem, Vogt observed was that "mosquito control that is intelligently conceived, expertly prosecuted, adequately maintained, and completely justified is as rare as the Eskimo curlew."[82]

Vogt disputed the claim that mosquito control had raised property values. He characterized Bishopp and Headlee's "annually reiterated" claim that pest mosquito control had raised property values "as an advertising man's technique." Neither Bishopp nor Headlee, Vogt observed, had ever addressed the fact that there had been "similar increases" in property values in places such as Maine, the Adirondacks, and White Mountains where there had never been any mosquito control. Property values had risen all along the East Coast.[83]

Vogt charged that the federal mosquito control projects had added in many places to the existing mosquito problem. He based this unlikely sounding claim on his careful review of Headlee's and Bishopp's presentations before the NJMEA. Both men had stressed the importance of maintenance work. Bishopp and Headlee warned that, without care, mosquito control ditches could become breeding sites. Vogt noted that local commissions had slashed funding on maintenance. In Atlantic County, New Jersey, more than thirteen million feet of ditches had required cleaning and upkeep. Using Headlee's estimate that it cost one cent per foot on a four-year repair schedule, Atlantic County should be spending forty-three thousand dollars annually for maintenance. The most recent Atlantic County budget showed that only four thousand dollars was spent on the ditches' upkeep. The net result was that the New Jersey Mosquito Extermination Commission, in the ditches taxpayers cannot afford to maintain, is breeding thirty-nine thousand dollars worth of mosquitoes every year.[84]

Vogt argued that the CCC work in Delaware presented a particularly flagrant example of the ineptitude and incompetence of the pest mosquito work. Citing data from the commission's reports, Vogt pointed out that between 1935 and 1936 there had been a 400 percent increase in the number of *Aedes sollicitans* caught in New Jersey light traps. During this period the Delaware Mosquito Control Commission had spent seventy-eight thousand dollars on mosquito control and "managed to treble their salt-marsh crop." Vogt discounted the alibis of the mosquito warriors who sought to "explain away several billion mosquitoes in a ditched county by saying there had been heavy rains or high tides. Now those who spend your money and mine destroying marshes and swamps have been able to get the cooperation of Mr. Harry Hopkins to the tune of many millions, but will they be able to get similar cooperation from that even Higher Power who controls marine and celestial waters." Vogt's conclusion was simple and direct. "*Something*," he declared, "*is damnably wrong with mosquito control*" [emphasis in original].[85]

The discussion that followed the formal presentations devolved into a shouting match. Bishopp accused Cottam of making such "free translations" of his published statements that "I am unable to recognize them." Bishopp claimed that Vogt had intentionally misinterpreted his offhanded observation: "what difference does it make if these birds are wiped out? There are more of them in Florida, aren't there?" Bishopp explained that when he had said this he had been explaining how "when mosquito control work drives wildlife out of a given area adjacent to a populous center it has the beneficial effect of protecting that wildlife by removing it farther from the murderous gun of the hunter, and I repeat that statement."[86]

The Third North American Wildlife Conference revealed the depth of the division between the nascent wildlife movement and the mosquito crusade. In some ways, this dispute reflected a clash between different generations. Bishopp had joined the USDA's Bureau of Entomology in 1904. Williams was a veteran of years of anti-malaria work in the south. Cottam and Vogt represented a new generation. They had little experience with malaria and the broods of salt marsh mosquitoes that plagued both the East and West coasts at the beginning of the twentieth century.

Bishopp and Williams left the debate feeling abused and misunderstood. They believed that Cottam and Vogt had intentionally misrepresented their positions. Both men had warned in 1933 at the beginning of the CWA initiative that it was essential that the federal mosquito work be competently planned and maintained. Cottam and Vogt dismissed these statements as insincere.

It is unclear what Bishopp and Williams had expected from the debate. Vogt's 1937 *Thirst on the Land* pamphlet had presented the case against mosquito control. It is surprising that Bishopp and Williams had not prepared detailed answers to Vogt and Cottam's arguments. The debate left both men concerned over the future of the anti-mosquito movement.

The Third North American Wildlife Conference marked the close of the first phase of the mosquito crusade. The era that began with John Smith's call for ditching the New Jersey salt marshes came to its end in the late 1930s. The first generation of mosquito warriors relied on shovels, ditching machines, and oilcans in their fight against mosquitoes. In the 1940s a new generation of mosquito fighters would improvise new chemical tools that opened new possibilities for mosquito control.

After the North American Wildlife Conference, William Vogt took up the larger issue of overpopulation. Vogt became convinced that the growing environmental problems were linked to population pressure. In 1951 Vogt became national director of Planned Parenthood Federation. Like Will Rogers, he believed that pamphlets on birth control should be widely distributed. Vogt differed only in that he thought they should be given to humans. Mosquitoes could take care of themselves.[87]

Chapter 7 Weapons of Mass Destruction

EIGHT DAYS AFTER THE Japanese attack on Pearl Harbor, Hawaii, the California Mosquito Control Association held its annual meeting. The proceedings convey a sense of urgency. The meeting opened with a symposium on encephalitis. In the discussion that followed, Morris Stewart, a researcher at the University of California's George Hooper Foundation, noted the value of the Hooper Foundation's ongoing investigation of encephalitis in the Yakima Valley in Washington State "in view of the present emergency." "In most [mosquito control] districts," Stewart warned, "we have conveniently classified certain mosquitoes as of primary importance and we have devoted our attention almost exclusively to those. Now we may be faced with the control of different species, and that means we must re-equip ourselves with the mental elasticity we had when we first went into this work."[1] Seasoned mosquito fighters like Herms and Freeborn advised their younger colleagues to prepare for change. "There are," Stanley Freeborn observed, "interesting days ahead for all of us. We know more about mosquitoes than we did at the time of the last war, but we need to know a great deal more."[2]

The anti-mosquito movement underwent profound changes in the 1940s. Concerns about wildlife issues faded as war clouds gathered over the Pacific and in Europe. New insect-borne diseases emerged to threaten both civilians and the armed forces. After Pearl Harbor, men like Herms, Freeborn, Bishopp, and Williams led efforts to protect soldiers, sailors, marines, and war workers from insect-borne diseases. In laboratories in Orlando, Florida, and Beltsville, Maryland, USDA researchers pioneered innovative means of chemical control while scientists and physicians at the Rockefeller Foundation and engineers and entomologists at the Tennessee Valley Authority (TVA) explored new strategies to reduce malaria parasites and control disease vectors. Simultaneously, hundreds of young soldiers and sailors learned the basics of mosquito control. Finally, as Allied Forces were advancing on Tokyo and Berlin, Robert Glasgow, Don Rees, Thomas Headlee, and Tommy Mulhern organized a national mosquito control association with the goal of consolidating the wartime accomplishments and catalyzing the movement's expansion in the postwar era.

Preparations for America's entry into the war began in 1939. One week after the Nazi invasion of Poland, Franklin Roosevelt declared a limited national emergency and called Congress into a special session. In the days that followed, Roosevelt won a series of measures that placed the country on a war footing. The House and Senate voted to repeal the 1935 Neutrality Act that prohibited the sale of arms and armaments to warring nations. In 1940 Congress approved Roosevelt's request for a staggering seventeen-billion-dollar defense budget. In September 1940 Congress authorized the first peacetime draft in American history. By Thanksgiving the first wave of the sixteen million men between the ages of twenty-one and thirty-five submitted their registration forms to their local Selective Service Board.

Protecting recruits and inductees from mosquito-borne diseases posed a huge challenge. Despite the U.S. Public Health Service's (USPHS) efforts during World War I, there had been more than ten thousand cases of malaria in the military.[3] Dr. Joseph Mountin, director of the USPHS's State Service Division in Washington, made establishing a comprehensive malaria control program a priority. In 1940 Mountin sent Louis Williams to serve as the USPHS's liaison with the Fourth Army Corps headquarters in Atlanta.[4] Williams, who began his career in malaria control in World War I, had recently returned to the United States from Southeast Asia where he had served as chief of the Malaria Commission for the China Burma Highway project. His experiences in Asia strengthened his conviction in the importance of malaria control for the war effort.

Historically, malaria had shown a steady decline in the United States since the midpoint of the nineteenth century. The fact that malaria's retreat from the Northeast and Midwest began before Ross's discovery of the role of mosquitoes as the disease's vector has led some historians to minimize mosquito control's role in suppressing the disease. Darwin Stapleton, executive director of the Rockefeller Archive Center, observed that the "lessons of malaria control in the twentieth century remain ambiguous."[5] In 1994 J. de Zulueta argued in *Parassitologia* that "the migration of the rural population, the main support of malaria in past times, to urban areas, the improvement of housing conditions of those who remained in the country, and last but not least, the general improvement of the social and economic conditions led to the end of malaria in practically all industrialized countries."[6] Margaret Humphreys developed this thesis in her 2001 book, *Malaria: Poverty, Race, and Public Health.* Humphreys attributes malaria's decline primarily to the migration of African Americans out of the rural South.[7]

Zulueta's and Humphreys's revisionist interpretation merits serious consideration. No one, however, at the USPHS or the War Department in 1940 could predict how great a threat malaria would pose to the war effort. Like most epidemiologists in the 1930s, Louis Williams believed that the disease moved in seven-year cycles. The current cycle reached its high point in 1934. Since then the malaria rate had fallen throughout the Southeast. There were, however, disturbing signs of an increase in cases along the South's infamous "malaria belt." Williams feared that the disease's next "upswing had already commenced."[8] If true, then 50 percent of the army's new recruits were at risk.

When Williams arrived in Atlanta, he immediately set about organizing mosquito control on the bases throughout the Fourth Service Command. His first priority was recruiting entomologists and engineers to supervise the work. Williams relied on Henry Hanson, Mark Boyd, John Elmendorf, and Willard King for assistance. Hanson, who served two terms as Florida's chief health officer, had made Florida a center for anti-malaria research. In 1932 Hanson had persuaded Mark Boyd, a senior malariologist and former colleague at the International Health Board, to establish the Rockefeller Foundation's Malaria Treatment Station in Tallahassee. Two years earlier Hanson convinced the USDA to transfer Willard King and the Bureau of Entomology's mosquito research station from Mound, Louisiana, to Orlando, Florida.

Hanson stepped down as Florida's health officer in 1935. The effort to make Florida a center for mosquito research and control work continued. In 1937 the State Health Board and the Rockefeller Foundation agreed to cosponsor a malaria demonstration project in Escambia County. The foundation assigned John Elmendorf, a distinguished epidemiologist, to direct the project. John Mulrennan, Florida's state entomologist, and David Lee, the state's sanitary engineer, served as Elmendorf's lieutenants. In April 1940 Gordon Fair, a professor at the Harvard School of Public Health, reviewed the project. Fair told a local newspaper reporter that the Florida Escambia Project "was the most complete work on malaria control that I have ever seen. It proves how engineering principles can be applied to a sanitation problem."[9]

Williams patterned his plan for the military anti-malaria work on the Florida program. During the "national emergency" USPHS entomologists and engineers would supervise Army and WPA workers' anti-malaria control efforts within military facilities. Individual states were assigned responsibility for the civilian areas surrounding the bases. USPHS teams set to work at military sites in April. In July 1941 the Florida State Board of Health formalized its relationship with the USPHS and named John Elmendorf the head of the state's Bureau of Malaria Control, which became a model for the other states in the Fourth Service Command. By year's end, USPHS entomologists and engineers had guided WPA workers in establishing anti-malaria control programs in twenty-seven training facilities in Tennessee, North Carolina, South Carolina, Mississippi, Alabama, Georgia, and Florida.[10]

In the months leading up to Pearl Harbor, concern about the threat of insect-borne diseases was not limited to the Fourth Service Command. In 1940 Chester Gillespie, chief of the California Health Department's Bureau of Sanitary Engineering, hired twenty-five-year-old entomologist Richard Peters to lead the engineering bureau's mosquito control program. Peters, a UC Berkeley graduate, was part of a cadre of extraordinary students whom Herms trained in the 1930s.

It would be difficult to find a more impressive group of apprentice entomologists. Herms's students included George "Ned" Bohart, Thomas Aitken, Paul DeBach, William Reeves, Robert Usinger, and George Ferguson. These future entomological giants were also formidable athletes.[11] In 1938 Peters led the entomology department's basketball team to the UC intramural championship game. Peters and Ned Bohart, who became an expert on pollinating insects, were the guards. Paul DeBach, who later won international acclaim for his work on biological control at UC Riverside, was the team's center. Thomas Aitken, who was already a graduate student and would go on to a distinguished career at the Rockefeller Foundation and Yale, was one of the forwards. William Reeves, who had transferred to Berkeley from Riverside, was the other forward.[12] Herm's all-stars remained undefeated until the championship game. They might have prevailed if "Billy Bugs" Reeves had not been sidelined with a torn medial collateral ligament.[13]

Peters, who graduated in 1938, found work with the California Department of Agriculture in a series of short-term jobs ranging from potato inspector to border quarantine agent. A year later he left the agriculture department and joined the Monterey Health Department. There he served as a general assistant to the county health office. In practical terms, this meant that Peters divided his time between sanitation, epidemiology, and mosquito control. During an inspection visit to Monterey, Chester Gillespie, head of the Bureau of Sanitary Engineering at the State Health Department, was surprised to learn that the tall, sandy-haired worker he had observed spraying oil had a degree in entomology. When some of his staff protested Peters's appointment because he did not have an engineering degree, Gillespie created the post of state mosquito control officer.[14]

Peters had a twofold assignment. First, Gillespie wanted him to assist the state's eighteen mosquito districts. Unlike New Jersey, where the agricultural experiment station guided the state's mosquito commissions, California lacked a central agency to coordinate the state's anti-mosquito work. There was a tremendous disparity between the various mosquito programs. Some districts like Alameda, where Harold Gray had served as superintendent since 1930, were well organized and ably led. Others were not as fortunate. Gillespie assigned Peters the task of being a "hand-holder" for the mosquito abatement districts (MAD). The job required him to meet with the superintendents, review current work, and collect mosquito pools, which he would bring to Berkeley for identification. Aitken and Reeves served as Peters's taxonomists. When Aikens and Reeves had classified the mosquito pools, Peters returned to the MADs and briefed the superintendants on the results "asking them whether they had tried this or had tried that, and observing their problem areas."[15]

Peters's responsibilities expanded in 1941 after Gillespie received a telegram from Washington designating his agency the "maternal" sponsor for the statewide emergency National Defense WPA Mosquito Control Project. Gillespie named Peters "godfather" for the state health board's federally funded anti-mosquito defense work. By the December 1941 CMCA meeting, Peters had organized mosquito control programs at twenty-one of the state's thirty-five military establishments.[16]

Peters's Berkeley visits kept him in touch with the entomology department. From Reeves he learned of the Hooper Foundation's research on encephalitis in the Yakima Valley in Washington State. Reeves believed that *Culex tarsalis* was responsible for transmitting both the western equine encephalitis (WEE) virus and the St. Louis encephalitis (SLE) virus.[17] If Reeves was correct, then California and the rest of the nation faced a significant health threat from the mosquitoes that transmitted these pathogens. Peters took Reeves's findings to his superiors and convinced them that "encephalitis should be of equal status to malaria in eligibility for federal assistance" in the wartime effort.[18]

Serendipity, a basketball injury, and professional misfortune led Reeves to study mosquitoes and encephalitis. Born in 1916, Reeves grew up on a ranch near Riverside, California, where his father was a beekeeper. As a boy he described himself as an indifferent student more interested in chasing bugs and playing ball than his studies. After graduating from high school Reeves spent two years at Riverside Junior College.

Reeves took his first step toward his eventual career during this period. After his freshman year, Reeves sought summer work at the University of

California research station in Riverside but discovered that there were no jobs for junior college students. In desperation, he appealed to the station's director, Henry Quayle. Quayle, who had become one of the leading authorities on citrus and the biological control of insect pests after his foray into mosquito control in Burlingame, told Reeves there were no jobs. Reeves persisted, offering to work for nothing. Finally, Quayle relented and gave Reeves a job at fifty cents an hour mounting scale insects on microscope slides.

At Riverside, Reeves had his first taste of research. Lunchtime conversations with Quayle and other professors such as Harry Smith, Harold Compere, and P. H. Timberlake encouraged Reeves to pursue his interest in entomology. He transferred to Berkeley in 1936. Decades later Reeves maintained that his injury in the basketball play-offs was the turning point. The torn cartilage "was the best thing that ever happened to me, because I could no longer play basketball . . . I started studying."[19]

Unlike Peters, Reeves decided to remain at Berkeley and pursue a doctorate when he graduated in 1938. Herms arranged a research assistantship in the entomology department as well as a job working for Harold Gray at the Alameda County MAD. Gray made Reeves responsible for answering complaints caused by *Aedes sierrensis* (formerly known as *Aedes varipalpus*). *Ae. sierrensis* is a small black mosquito with white legs that lays its eggs in tree holes. When the tree holes fill with water, the eggs hatch. Reeves's job was to find the tree holes, climb the trees, and kill the larvae.

Curiosity led Reeves to start a colony of the *Ae. sierrensis* mosquitoes. From Herms he learned that little was known about the insect's history. Reeves decided to study the mosquito. He thought it might make his job easier and supply him with a dissertation topic. After a couple of false starts, he succeeded in establishing an *Ae. sierrensis* colony.

A persistent problem plagued his work with the tree-hole mosquitoes. No matter what Reeves did, he could not get the eggs to hatch in numbers sufficient for his research needs. A chance conversation with another graduate student working in bacteriology suggested an experiment. Reeves placed a culture of whooping cough bacteria in the water and all of the eggs hatched. More experiments followed. Reeves discovered that it was not the bacteria that caused the eggs to hatch but rather the depletion of the water's oxygen level that triggered the hatching. No one had made the connection between oxygen and the hatching process. Reeves believed he could use these experiments as the basis for his thesis.[20]

Initially, all went well. Herms and Freeborn approved the topic and suggested he report his preliminary findings at the December CMCA meeting. A few weeks later Herms asked him to show Fred Bishopp his *Ae. sierrensis* colony.

Bishopp had stopped in Berkeley on his way to visit the USDA's research station in Portland, Oregon. Herms had mentioned Reeves's dissertation, and Bishopp had asked to hear more. Reeves was in a quandary. He knew that USDA researchers in Oregon were also working on tree-hole mosquitoes. He was reluctant to reveal his findings. His hesitancy in answering Bishopp's questions angered Herms. Finally, Herms ordered Reeves to give Bishopp a detailed explanation of his research. He had no option but to give Bishopp his protocols.[21]

Months passed and Reeves heard no more about the visit. A few days before the CMCA meeting, Harold Gray phoned Reeves. Gray, the program chair, told Reeves that Claude Gjullin, one the USDA researchers at the Oregon field station, had asked for a few minutes provide an update on his research. At the meeting, Reeves sat quietly as Gjullin described his discovery that lowering the oxygen level caused tree-hole mosquito eggs to hatch. When Gjullin finished, Gray invited Reeves to the lectern to read his paper on *Ae. sierrensis*. Reeves rose and declared, "I don't need to waste your time. Mr. Gjullin gave my paper for me."[22]

Gray called a recess, and Herms and Freeborn rushed to the front of the room demanding to know what happened. They discovered that when Bishopp had reached Portland, he suggested that Gjullin adopt Reeves's protocols. Bishopp had not mentioned his visit to Berkeley. Gjullin was innocent of any wrongdoing. Herms and Reeves were furious. Herms phoned Bishopp and demanded an explanation. Herms reported that Bishopp had laughed and said, "That's life."[23]

Fred Bishopp cost Reeves his dissertation. He never apologized. Bishopp's stop in Berkeley and subsequent visit to Oregon was tied to the war effort. Early in 1940 representatives of the Army, Navy, USPHS, and USDA had met to consider "the disease-transmitting potentialities of insects in a global conflict."[24] In May the Army's surgeon general, James Magee, sent a formal request to the National Research Council for guidance in developing a plan to protect American troops against insect-borne diseases. The National Research Council's Division of Medical Sciences established two subcommittees to study the problem: the Committee on Chemotherapeutic and Other Agents and the Subcommittee on Tropical Diseases. Bishopp's trip to the West Coast grew out of the Subcommittee on Tropical Medicine's effort to prepare a strategy for research on repellents, insecticides and other anti-mosquito measures. He hoped to learn three things in his trip to Oregon. First, he wanted Harry Stage's assessment of the mosquito control issues facing the Pacific Northwest in the national emergency. Second, he wanted Edward J. Knipling's advice on insecticides and repellents. And finally, he wanted an update from Gjullin on the field station's research projects.

Interest in mosquito control in the Pacific Northwest had surfaced in the late 1920s when H. H. Riddell, a prominent Portland attorney, had written to the USPHS. Riddell wanted know what he could do about "hordes of mosquitoes" (*Aedes vexans*) that plagued his summer retreat on the Columbia River in Skamania County, Washington. Eventually Riddell's letter made its way to Harrison Dyar at the Smithsonian Museum. Dyar, who had visited the Columbia River Valley on a collecting trip, wrote to Riddell and advised him "to inspect the water at a certain bend in the Columbia River . . . and he would probably find mosquito larvae." Riddell followed Dyar's instructions and "soon located one of the thousands of mosquito breeding areas in the lower Columbia River Valley."[25]

Coincidentally, Riddell learned that Portland's chamber of commerce had organized a meeting to discuss the city's mosquito problem. Riddell attended and described what he had learned from Dyar. At the meeting's conclusion, the chamber voted to form a mosquito control committee made up of representatives from Multnomah County, the City of Portland, and Oregon State's agricultural experiment station. The committee recommended that the chamber ask for the Bureau of Entomology's assistance in developing a mosquito control plan. George Bradley made a preliminary survey of the Portland area in 1929, and a year later Willard King completed the survey.

Harry Stage was the Bureau of Entomology's expert on the mosquitoes of the Pacific Northwest. Stage had established a USDA field station in Portland in 1930. Four years later Bishopp placed Stage in charge of the CWA and WPA pest mosquito projects in the Pacific Northwest. Later, Bishopp assigned Edward Knipling to the station. Knipling, who joined the USDA in 1931 after earning his bachelor of science degree from Texas A&M University and his master's degree from Iowa State University, was one of the Bureau of Entomology's most promising researchers on insecticides and control strategies. Bishopp had sent Claude Gjullin to Portland to assist Stage in his research on the region's floodwater mosquitoes. In a 1934 report on region's floodwater mosquito problem, Stage argued that these mosquitoes "were much more than a nuisance." Floodwater mosquitoes "caused losses running into the hundreds of thousands of dollars through decrease in milk production in dairy herds, inability to hire or keep labor in truck crop area, and loss of tourist trade."[26]

Gjullin thought that the key to controlling Oregon's floodwater pests lay in understanding the mechanism that triggered the *Aedes* mosquitoes eggs to hatch. When Bishopp learned that Gjullin's research had stalled, he suggested that he adopt a new approach. He neglected, however, to tell Gjullin where the ideas had come from. Bishopp's insensitivity cost Reeves his thesis. Herms and Freeborn told him that he would have to find a new topic.

Several weeks after the CMCA debacle, Reeves approached Karl J. Meyer, who was the director of the George Hooper Foundation in San Francisco. Reeves had heard Meyer lecture in Herms's medical entomology course on viruses. Reeves explained his situation and told Meyer he was interested in learning more about the relationship of mosquitoes and viruses. At the time, there were no courses on virology at Berkeley. Meyer, who suspected that mosquitoes played a role in transmitting viruses, welcomed the young entomologist's interest in virology.

The death of six thousand horses in the San Joaquin Valley in 1930 triggered Meyer's study of encephalomyelitis. Meyer, who was one of the leading figures in the study of bacteriology, immunology, and parasite vectors, joined the University of California in 1913 as an associate professor in the newly formed George Hooper Foundation for Medical Research.[27] Eight years later Meyer became the Hooper Foundation's director. In 1931 Meyer isolated the WEE virus in a horse's brain.

Meyer's unraveling the mystery of the WEE virus is one of the great stories in the history of modern virology. When Meyers learned that thousands of horses had died, he dispatched a laboratory technician to obtain brain tissue from horses that had succumbed to the mysterious disease. The technician failed to find an uncontaminated specimen for Meyer. Meyer learned from a UC Davis veterinarian, C. M. Haring, that there was a sick horse that met his specifications nearby in Merced County. Meyer decided to go to the Central Valley, find the horse, and bring its brain back to San Francisco for testing.

Years later Bill Reeves recounted the details of Meyer's discovery of the WEE virus. Meyer told Reeves that he and Haring had no trouble finding a sick horse. The problem was that the old farmer who owned the animal was unwilling to sell it. Meyer offered to pay him fifty dollars. The old man was adamant. As he was leaving, the farmer's wife pulled him aside and said she would sell him the horse. Together they devised a plan to outwit her husband. Meyer and Haring were to hide in the bushes along the roadside. When the old man was asleep, the woman would put a lamp in the window letting Meyer know the coast was clear.

The plan went like clockwork. The lantern appeared in the farmhouse window, and Meyer and Haring rushed to barn. They "knocked the horse out and chopped off its head." Fearing the farmer might awaken, the two men threw the horse's head in the backseat of the car and headed down Highway 99. At midnight they stopped at an all-night gas station. The sleepy gas station attendant watched in amazement as Meyer removed the animal's brain and packed it in ice. "By daybreak the next morning," Reeves reported, Meyer was back in San Francisco "and had a brain suspension inoculated into rabbits and guinea pigs, and they isolated the virus." Despite Reeves's persistent effort, Meyer would not reveal where the farmhouse was. Reeves always wondered what the old man thought when he went out to the barn and found his headless horse.[28]

The western epizootic and Meyer's successful isolation of the WEE virus was followed in 1933 with outbreaks of two other previously unknown forms of encephalomyelitis in the Northeast and Midwest. In New Jersey, Delaware, and Virginia, veterinarians reported the deaths of large numbers of horses from eastern equine encephalomyelitis. A few weeks later in August health workers in St. Louis reported the first of more than 1,000 human cases of a new disease that eventually produced 201 fatalities.[29] During the next five years, equine deaths from encephalitis continued to rise. Between 1935 and 1938 veterinarians estimated that more than 385,000 horses perished.[30] By 1941 there was evidence of the presence of the WEE, SLE, and EEE viruses in "all states from the West Coast east to Alabama and in the South and Michigan in the north."[31]

The appearance of three mysterious viral diseases between 1930 and 1933 sparked an intensive research effort. There was circumstantial evidence that mosquitoes were involved in transmitting the disease. In New Jersey, Headlee noted that all of the cases of the disease occurred in areas near salt marshes. In 1933 R. A. Kelser, an army medical researcher, reported his success in using an *Ae. aegypti* mosquito to infect laboratory animals with the WEE virus.[32]

Others researchers failed in their efforts to incriminate mosquitoes in spreading the diseases. In Maryland, Bishopp made the first attempt to capture infected mosquitoes in the wild. None of Bishopp's specimens was positive. In California, Herms failed in his effort to reproduce Kelser's findings. Further doubt was cast on the mosquito hypothesis when USPHS researchers issued their report on the St. Louis event. In tests relying on convict volunteers, USPHS epidemiologists reported the failure of mosquitoes to transmit encephalitis to human subjects. "As a result of this test," Herms erroneously declared, "mosquitoes apparently play no part in the transmission of encephalitis. The work will still go on, however, and where the least suspicion is cast upon mosquitoes, it should encourage us to greater and more careful research. We must eventually come off victorious against these pests which may take human life."[33]

Meyer, however, was convinced mosquitoes were involved in spreading encephalomyelitis. He named Beatrice Howitt to supervise field research of human and equine cases of encephalomyelitis in Bakersfield in 1937. Three years later he recruited William Hammon, a former missionary and Harvard-trained epidemiologist, to lead the Hooper Foundation's research on the disease. While Hammon and his family were in transit from Cambridge to Berkeley, Meyer received an urgent appeal for assistance from the Washington State Health Department. An encephalomyelitis epidemic had broken out in the Yakima Valley. Meyer phoned Hammon and told him to go directly to Yakima and begin an investigation.[34] In the summer of 1941 Bill Reeves joined Hammon in Yakima.

Reeves's assignment was to find out if arthropods transmitted the disease. Accomplishing this goal required collecting large numbers of insects for laboratory tests. Yakima Valley farmers grew accustomed to the tall young man traipsing through their pastures and prowling in their barnyards. It was a labor-intensive effort. To catch mosquitoes Reeves employed a mouth aspirator that he and Tommy Aitken improvised at Berkeley. The aspirator, a plastic tube with a copper mesh across its bottom, allowed Reeves to suck mosquitoes from the livestock. The first summer at Yakima he collected 12,466 mosquitoes feeding on cows and horses. His only selection criterion was that the horses and cows did not kick. Some of the animals grew to enjoy Reeves's personalized mosquito control service. Decades later, Reeves recalled "a big old bull that weighted about two thousand pounds . . . that just loved to have me collect mosquitoes off of him."[35]

Reeves and Hammon resolved the mystery of human and equine encephalomyelitis in Yakima. Their mosquito pools revealed the presence of both the WEE and SLE virus in *Cx. tarsalis* mosquitoes. Further experiments demonstrated that this species of mosquito could transmit encephalitis to laboratory animals. The discovery that *Cx. tarsalis* carried the virus in the wild was noteworthy. Entomologists had previously considered *Cx. tarsalis* unimportant as either a pest or a disease mosquito. Reeves's subsequent identification of high antibody levels against the virus in blood samples from the local wild bird population suggested a disturbing transmission pattern. Birds served as reservoir for the virus. Under the right conditions, *Cx. tarsalis* mosquitoes would feed on wild birds and, after an incubation period, transmit the virus to human beings.[36]

Reeves and Hammon's work attracted national attention. The *Newark Sunday Ledger* applauded their discovery. "One of the mysteries of medical science," the *Ledger's* reporter declared, "was believed solved last night with a government announcement that mosquitoes carry sleeping sickness." The editors of *Mosquito News*, the Eastern Association of Mosquito Control Workers' new publication, noted the discovery's significance for the war effort. "If these facts are established," the news article concluded, "a plan for mosquito eradication can be mapped that might control sleeping sickness as effectively as mosquito eradication helped control yellow fever."[37]

Reports from the Pacific in the weeks after Pearl Harbor underscored the significance of Reeves and Hammon's discovery. Controlling insect-borne diseases was essential. This became clear when, three days after the Pearl Harbor attack, the Fourteenth Imperial Japanese Army invaded the Philippines. On December 23, 1941, Gen. Douglas MacArthur ordered a retreat from Manila across the bay to the Bataan peninsula. Bataan's jungles and mountainous territory offered an ideal defensive position for the evacuees. Unfortunately, Bataan proved an equally marvelous habitat for anopheline mosquitoes. By early January army physicians reported cases of both *vivax* (benign) and *falciparum* (malignant) malaria. Soon 80 percent of the frontline units suffered from malaria. Supplies of quinine dwindled. By mid-January doctors cut the daily dosage of 10 grains of quinine in half. A month later only hospitalized patients received quinine. During the final desperate days before the "Battling Bastards of Bataan's" surrender, Japanese forces swept across British Malaya and captured Java, cutting off Allied access to 90 percent of the world's quinine production.[38]

The Allies and their Japanese adversaries encountered "a third combatant" in the opening weeks of the war. Malaria, Albert Cowdrey observed in his history of military medicine in World War II, wasted hundreds of thousands of Japanese and Allied combatants. Other mosquito-borne diseases such as dengue fever extracted a heavy toll. Cowdrey maintained that mosquito control contributed much to the Allied victory in the Pacific. Mosquitoes were indiscriminant in their search for blood. However, they delivered their "harshest blow to the army that could not replenish itself. 'General Disease' would surely be a factor in the Pacific war as 'General Winter' and 'General Mud' in the battles then raging on the plains of Russia."[39]

"General Malaria" and "General Dengue" waged war on both Americans and Japanese in the Pacific. The experiences of the First Marine Division on Guadalcanal underscored the importance of anti-malaria measures. The battle for Guadalcanal during the summer and fall of 1942 was nearly lost because of a lack of "malaria discipline." During the first weeks of battle, "malaria control was farcical." Before landing, many marines had thrown away their mosquito nets. Once on the island, those who had retained their anti-mosquito gear "had neither the time nor the strength to bother about mosquito bars and head nets and gloves."[40]

The results were predictable. Navy corpsmen reported that there were 900 cases of malaria in August 1942. By October the number soared to 2,630. In November a hastily organized malaria control unit launched a counter offensive. By February 1943 the anti-mosquito work of James Douglas, one of Karl J. Meyer's parasitologists at the Hooper Foundation, and a team of Navy medical entomologists showed positive results.[41] The toll of malaria, however, was staggering. More than sixty thousand servicemen were infected with malaria on Guadalcanal. The sobering lesson was not lost on Gen. Douglas MacArthur. "It would be," MacArthur observed in a conversation with Paul Russell, a veteran Rockefeller Foundation researcher, "a very long war indeed if every division facing the Japanese must count on a second division in the hospital with acute malaria, and a third division in a convalescent depot with relapsing malaria." If victory was to be achieved, the Allies must win the battle against insect-borne diseases.[42]

Between 1941 and 1945, American entomologists, chemists, toxicologists, and physicians launched a counteroffensive against mosquitoes and insect-borne diseases. Veteran mosquito warriors such as Willard King led the campaign in the Pacific. King, who had directed the USDA Orlando laboratory since 1932, received a commission as a lieutenant colonel in the U.S. Army Sanitary Corps in 1941. After a short stay at Fort McPherson in Georgia where he guided the development of training courses for military personnel on mosquito control, King received orders for the South Pacific in 1942.[43]

From his base at Port Moresby in New Guinea, King organized anti-mosquito work in the South Pacific. Dick Peters joined King's mosquito brigade early in 1943. Peters recalled that his first assignment was to "dig his own slit trench." Soon Peters was leading mosquito survey teams in combat with a "tommygun in one hand and a dipper in the other."[44] By the war's end, numerous University of California entomologists such as James Douglas, Stan Bailey, and Robert Usinger had joined King and Peters in the Pacific. Stan Bailey, who received his doctorate in 1931 at Berkeley and went on to chair the UC Davis entomology department, volunteered for active duty in the U.S. Navy medical corps shortly after Pearl Harbor. Bailey's leadership in the fight against dengue fever in the Central Pacific earned him the reputation of being the "savior of Guam."[45]

On the home front, Louis Williams Jr. and Stan Freeborn organized the effort to protect military personnel and war workers from the malaria menace. In February 1942 Thomas Parran, U.S. surgeon general, asked Williams to take charge of all national defense malaria control activities. Williams named Mark Hollis, a USPHS engineer, to be his executive officer. Stan Freeborn agreed to serve as the agency's senior malariologist.[46] In April, Surgeon General Parran officially changed the organization's name to Malaria Control in War Areas (MCWA).

Initially the MCWA's mission was to protect military personnel and war workers from malaria in fifteen southern states. In June naval officials, fearing outbreaks of dengue and yellow fever, requested that MCWA undertake

Aedes aegypti–control measures in Key West, Charleston, Miami, and other cities along the Gulf Coast. By 1943 the MCWA had grown to include twenty states, the District of Columbia, Puerto Rico, and the Hawaiian territory. The MCWA reached its high point in 1944 when nearly four thousand MCWA workers conducted control operations at 2,067 military establishments in 917 war areas.[47]

Florida was one of the first states to start MCWA work. Shortly after receiving Surgeon General Parran's directive, Williams contacted John Mulrennan, Florida's state entomologist, and requested that he organize a pilot MCWA program in Jacksonville. The goal was simple: "Protect war workers from mosquito-borne diseases."[48] Mulrennan had begun his career working for Mark Boyd at the Rockefeller Malaria Treatment Station. Elmendorf, a veteran Rockefeller scientist, supplied medical and technical expertise. Williams would supply the funds for the work. By July Mulrennan's and Elmendorf's MCWA teams had already applied 23,635 gallons of oil and 8,511 pounds of Paris green at thirty-eight military sites in eight war areas.[49] By year's end, MCWA workers had used 1,875,000 gallons of oil and 152,000 pounds of Paris green.[50]

Most MCWA officials considered the program a stunning success. The MCWA relied on mosquito counts to demonstrate their progress in vanquishing malaria. George Bradley reported that the seventy-eight MCWA entomologists had succeeded in "maintaining satisfactory low densities of anophelines at approximately ninety-two percent of over nine hundred war establishments" by the end of 1942.[51] Overall, "the number of cases of malaria acquired in any one of the protected areas has in the great majority of instances been zero, and in the remainder of the areas, the cases have been so few and so scattered as to be practically negligible."[52]

Stan Freeborn was skeptical about the results. Since its high point in 1935, the malaria rate had fallen each year. Certainly the MCWA had contributed to the effort to protect servicemen and war workers. Freeborn cautioned against taking too much credit. "I think," he declared, "some of us who are closest to the work sometimes confuse costs, size of projects, gallons of oil and mosquito densities as criteria of accomplishment whereas our only true measure is the presence, severity, reduction or absence of malaria. In the face of a declining malaria rate it would be presumptuous to take too much credit for our operations."[53]

No one questioned the significance of the research on insecticides, repellents, and development of malaria drug compounds. In October 1941 the National Research Council's Subcommittee on Tropical Medicine issued a report calling for the Office of Scientific Research and Development to fund a research program on insecticides and other means of preventing insect-borne diseases. The report recommended that the USDA's Bureau of Entomology and Plant Quarantine be given responsibility for the research effort.

In 1941 the secretary of the Department of Agriculture named Bishopp assistant chief of the Bureau of Entomology and placed him in charge of the bureau's research effort. Initially, Bishopp asked Willard King to oversee the insecticide and repellent research at the USDA's Orlando laboratory. King was called to active duty in the army sanitary corps before the work started. In July Bishopp chose Edward Knipling to lead the research team.

During the next thirty-six months USDA entomologists, toxicologists, chemists, and engineers tested more than ten thousand chemical compounds. Initially the primary objective of the lab was to conduct investigations in the control of body lice (the vector of typhus) and insect repellents (mosquitoes, biting flies, mites, fleas, ticks, and bedbugs.) Knipling's instructions were simple: His researchers were not to concern themselves with basic research. Their objective was to examine and quickly assess the value of various materials that might be of use against disease-bearing insects and pests.[54]

The goal was to find practical tools that could be used in combat. "We cut corners," Knipling recalled, "whenever possible to reach an end point and to make specific recommendations or to suggest what might be tried by the army in large-scale practical tests."[55] Fred Bishopp emphasized the need for speed. "Disease," he explained, "was a more formidable enemy than the Japs and that without the effective control of disease the outlook for early defeat of the Japanese was indeed gloomy."[56]

Late in 1942 the Orlando researchers made a revolutionary breakthrough. In November Bishopp forwarded to Knipling samples of an insecticide he had received from the Geigy Chemical Company of New York. Paul Mueller had discovered the chemical, called Gesarol, in Geigy's Swiss laboratories in his search for a potential tool against moths and lice. USDA chemists in Orlando identified the active agent in Gesarol as the chemical compound dichlorodiphenyltrichloroethane. Knipling's researchers followed the lead of British chemists and used the initials DDT to designate the chemical agent.

The discovery of the insecticidal properties of DDT marked the beginning of a new epoch in mosquito control. Since John Smith's pioneering work in the New Jersey meadowlands and marshes, ditching and drainage had served as the principal strategy for mosquito control. Chemical means of control were limited to oil, Paris green (after 1921), and pyrethrum (after 1930) as larvicides. DDT's low cost, potency, persistence, and efficacy as both a larvicide and adulticide revolutionized mosquito control.[57]

Field tests in early 1943 confirmed DDT's potency against lice, mosquito larvae, and adult mosquitoes. In May 1943 the Army Quartermaster Corps added Neocide, the Orlando laboratory's formulation of a DDT louse powder, to its stock of pesticides. The first shipments of Neocide reached North Africa in the late summer. Fred Soper, a Rockefeller Foundation scientist who had relied on Paris green in the 1930s to eradicate *Anopheles gambiae* in Brazil, was one of the first to use Neocide in delousing German prisoners in North Africa.[58] Neocide's power lay in its persistence. Other insecticides required repeated treatments. A single application of Neocide lasted for months.

The discovery of DDT's power as a lousicide sparked a dramatic expansion of the Orlando laboratory's mission. In July 1943 Knipling redefined the laboratory's mission. Four of the laboratory's nineteen researchers would continue work on lice while the remainder would be assigned to three teams. Chris Deonier led the effort to develop an effective mosquito larvicide. Arthur W. Lindquist was in charge of work on mosquito adulticides. The remaining investigators focused on refining Phillip Granett's research on insect repellents.

Deonier's and Lindquist's laboratory and field experiments demonstrated DDT's potency and persistence as both a larvicide and adulticide. In one test, a mixture as small as 1 part DDT per 100 million parts of water killed 100 percent of the test larvae.[59] Lindquist's findings were potentially even more important because they opened the possibility of an effective adulticide. In the early 1930s Joseph Ginsburg, a Rutgers biochemist, had achieved limited success killing adult mosquitoes using pyrethrum-based insecticides in hand pump spray guns. At the beginning of World War II, Flit (a pyrethrum-based commercial insecticide) was one of the most widely advertised of these products. Dr. Seuss drew a series of cartoons in a nationwide "Quick Henry, the Flit" advertising campaign.[60] Lindquist discovered that unlike pyrethrum, which rapidly loses its toxicity, DDT retained its potency. Experiments demonstrated that residues of DDT continued to kill adult mosquitoes months after the application.

In June 1943 at the Wilson Dam, TVA technicians sprayed DDT for the first time from an airplane.[61] Other aerial tests followed against larvae and adult mosquitoes at the Banana River Naval Air Station near Cocoa, Florida. In Stuttgart, Arkansas, James Gahan demonstrated DDT's phenomenal residual power in a series of tests in 1943 and 1944. Finally, in May 1945, the Orlando laboratory and the Rockefeller Foundation launched a major experiment in Mexico. Four months after the DDT aerial application there was nearly a complete eradication of adult mosquitoes within the treated buildings. The mosquito larvae in rice fields, which were adjacent to the village, dropped 90 percent.[62]

At Rutgers, entomologists voiced some of the earliest concern about DDT's impact on wildlife and the environment; Ginsburg called for restraint

Figure 7.1. DDT application to a U.S. serviceman. Photo courtesy of the Public Health Image Library, Centers for Disease Control and Prevention.

in using what became known as the "atomic bomb of the insect world."[63] Ginsburg, who developed the first mosquito adulticide, warned of DDT's potential harm. In July 1943 Knipling sent Ginsburg a sample of DDT for testing. Ginsburg conducted thirty-two field experiments using DDT as a larvicide in Essex, Middlesex, Monmouth, Passaic, and Union counties. He reported that "DDT, in concentrations sufficiently high to kill subsurface feeding larvae, was toxic to at least three species of fish." His conclusion was that "DDT may offer a new and highly potent weapon to supplement our present chemicals used in the control of mosquitoes. But before it can be recommended for practical use, more extensive field investigations under various mosquito breeding conditions are necessary."[64]

While Ginsburg voiced concern about DDT's effect on wildlife, Phillip Granett, another Rutgers researcher, began research to find a repellent that would be effective against *Ae. aegypti* mosquitoes. Granett tested more than one thousand different compounds.[65] By 1940 Granett had identified three promising candidates: Dimethyl phthalate, Sta-Way, and Rutgers 6–12.

The Orlando group adopted Grannet's approach to testing repellents. Instead of focusing on *Ae. aegypti*, the Florida researchers focused on finding a repellent against the most important southern vector of malaria, *An. quadrimaculatus*.[66] The challenge facing the Orlando researchers was to formulate a repellent that would be effective across a wide range of insect pests. Eight thousand formulations were tested. A mixture consisting of 60 percent of dimethyl phthalate, 20 percent Indalone, and 20 percent Rutgers 6–12 proved most effective.[67] Following the Rutgers tradition of using the percentages of the active chemical agents present as the formulation's code name, the new repellent was designated 6–2-2. During the war years, the military purchased more than "one hundred and fifty million two-ounce bottles of synthetic organic repellents."[68]

Neither 6–2-2 nor any other of the repellents developed in Orlando or Rutgers were panaceas. At best, the repellents provided limited protection against only a few mosquito species. Often, "repellents . . . effective against a number of Florida species of mosquitoes and other insects did not produce the desired results with some species of the Tropics, *Anopheles farauti* in particular." For most soldiers, however, the compelling argument against the repellents was that they were unpleasant to wear. "Unless mosquitoes were biting in unendurable numbers," Emory Cushing concluded, "the soldiers were disinclined to use repellents. Some combat troops refused to use them on the grounds that in the humid areas the odor was detectable by the enemy." It was, they decided, far better to be bitten by a mosquito than killed by Japanese infantryman.[69] The search for an effective repellent continued in Orlando after the war's end. In 1954 USDA researchers in Orlando succeeded in formulating a broad spectrum repellent called DEET (N,N-diethyl-m-toluamide).

The research effort was not limited to repellents and insecticides. During the war USDA teams in Orlando and Beltsville, Maryland, devised new ways to broadcast insecticides. In Orlando, Knipling assigned Chet Husman and Olin Longcoy the task of enhancing existing airplane spray units. Husman, an engineer, and Longcoy, a pilot, came up with a number of modifications. Earlier efforts to use airplanes in mosquito control had failed because the "gallonage [of pesticide] required was so great."[70] DDT solved this problem. Field tests revealed that as little as two quarts of a DDT petroleum mixture gave near total control of a one-acre area.[71] Husman and Longcoy devised a breaker-bar apparatus that could be easily mounted on a number of different aircraft. Pilots at the Banana River Naval Air Station dubbed the Husman-Longcoy device a "Flying Flit Gun."

The "Flying Flit Gun" was one of a series of USDA innovations that revolutionized insect control. In 1940 a team of USDA scientists in Beltsville, Maryland, produced the first aerosol spray device. The effort to develop power sprayers began in 1907 in New Jersey with a device that used compressed air to "discharge the spray though a hose line and from the spray nozzle."[72] The fabrication of the aerosol spray can marked a significant advance in spray technology. W. N. Sullivan, L. D. Goodhue, and J. H. Fales began their examination of different methods of insecticide dispersal in the late 1930s. They discovered in early tests that a fine insecticidal mist was produced when insecticides were sprayed on a hot metal plate. Later Goodhue and Fales succeeded in creating a powerful insecticidal mist by mixing a pyrethrum concentrate, sesame seed oil, and Freon 12 in a metal canister.[73]

The Beltsville aerosol technology had wide applications. After the war, Freon-based aerosols would be used to spray deodorants and whipping cream. In 1943 toxicologists in Orlando created a DDT-pyrethrum spray formula for the aerosols. The DDT aerosol bombs "received widespread popularity with troops because of the convenience in application of insecticides and effectiveness. A few seconds spraying quickly rid quarters, tents, and dugouts of mosquitoes." By 1945 the military had purchased more than thirty-five million aerosol bombs.[74]

The first practical use of the aerosol bombs came in 1941, when the USPHS used Beltsville pyrethrum aerosols to disinfect airplanes on intercontinental lines.[75] In the 1920s T.H.D. Griffitts conducted a series of experiments that demonstrated that mosquitoes were frequent fliers on flights traveling between Cuba and other Caribbean islands and Miami.

Air travel, in fact, did lead to a major dengue outbreak in Hawaii in 1943. Aerosol bombs, while effective in killing mosquitoes, provided no protection against the movement of infected humans. The precise origin of the 1943 dengue epidemic is uncertain. It was widely believed that two airplane pilots carried the disease to Oahu in mid-July. On July 24 two individuals were hospitalized with dengue fever. Two weeks later physicians reported numerous cases in the Waikiki area.[76]

"Mosquitoes," Robert Usinger observed, "are recent immigrants to Hawaii." Usinger, who had been one of Herms's students and was then an MCWA entomologist assigned to the Honolulu dengue control effort, began his 1944 report on the dengue epidemic with a brief history of the winged pest in the "Hawaiian paradise." In 1826 a native Hawaiian told a missionary about a curious insect that was "singing" in his ear. It was later determined that *Culex quinquefasciatus* mosquitoes had traveled to Maui aboard the sailing vessel *Wellington*. Lacking a word for mosquito in Hawaiian, the natives created a variant on the English and called the singing pests "makikas." By 1900 *Aedes aegypti* and *Aedes albopictus* "makikas" had established themselves in the tropical paradise.[77]

Both *Ae. aegypti* and *Ae. albopictus* are competent vectors for dengue virus. In 1893 natives complained of "boo-hoo" sickness, whose symptoms resembled dengue fever. The first clinically diagnosed outbreak took place ten years later. Local physicians estimated that thirteen thousand of the islands' forty thousand residents were infected with the disease. Dengue outbreaks recurred annually from 1911 to 1915.[78]

The possibility of an "explosive" outbreak of dengue fever on Oahu prompted immediate action. "The effect of thousands of cases of dengue fever among military personnel and the war-connected civilian population in the principal base of the Pacific Ocean might have been disastrous," the director of the MCWA's Dengue Mosquito Control Board in Hawaii declared. On August 8 military officials declared downtown Honolulu and Waikiki off limits and assigned troops to support the Territorial Board of Health's Rat and Mosquito Control Squad.[79]

Mark Hollis, MCWA's director, dispatched a team of MCWA engineers and entomologists to coordinate the campaign to suppress the *Ae. aegypti* and *Ae. albopictus* populations. The unit recommended immediate action to prevent dengue's spread to the neighbor islands. Local health workers used Beltsville aerosol bombs to disinfect airplanes leaving and returning to the islands. Local radio stations broadcast daily bulletins on the anti-dengue campaign. Each of the territory's sixty movie theaters screened a two-minute film clip as part of the governor's "Work to Win" plan to stop the disease's spread. By June 1944 Usinger was confident the MCWA team had succeeded in containing the outbreak. MCWA epidemiologists reported that only a handful of the 1,498 dengue reported cases during the outbreak had involved military personnel.[80]

The Hawaii dengue outbreak added urgency to the ongoing debate about the organization and goals of the anti-mosquito movement. By 1943 there was widespread concern that returning veterans might introduce malaria and other tropical diseases into the civilian population. Don Rees reported an outbreak of malaria in Brigham City, Utah, near Bushnell General Hospital where returning veterans were receiving treatment. The problem, however, was not limited to military hospitals. Rees noted that there had been an increase in malaria in Ogden, Utah, near a recently opened prisoner-of-war camp.[81]

Congress grew alarmed about returning veterans spreading mosquito and insect-borne diseases in 1944. When Utah senator Elbert Thomas learned of the Brigham City malaria epidemic, he rallied congressional support for what became Public Law 410. Thomas told his colleagues that the law was needed because "there are coming back to our country from all parts of the world men afflicted with malaria and other sicknesses."[82]

Joseph Mountin, the assistant surgeon general at the USPHS, credited Louis Williams Jr. with being the driving force behind the effort to redefine the mission of the MCWA program. In 1943 Williams stepped down as head of the MCWA to accept an assignment as the U.S. Army's malariologist in the Mediterranean theater of operations. In North Africa and Italy, Williams witnessed DDT's insecticidal power. Conversations with Fred Soper, led to speculations about the postwar era. Soper recounted his success in eliminating *Anopheles gambiae* in Brazil. Williams was confident that the MCWA could be used to eliminate malaria in the United States. The goal of the original MCWA program had been to protect military personnel and war workers from infection in areas where malaria was endemic in the population. The new program represented a 180-degree shift in focus. Under the new directive, MCWA teams would seek to protect civilians from returning military personnel.[83]

The idea for the Extended Malaria Control Program took shape in 1944. A first step would be the formation of "a group of mobile units" strategically placed to respond to malaria outbreaks in areas near prisoner-of-war camps, military hospitals, and debarkation centers. These units would serve as "'circuit riders' spreading the gospel of anopheline control."[84]

The distinguishing feature of the USPHS Extended Malaria Control Program in 1945 was its reliance on DDT for residual control. The idea of residual control was to prevent an infected mosquito taking a second blood meal. If the mosquito could be killed, it would break the transmission cycle. Unwittingly, Stan Freeborn pioneered the idea of residual control at his duty station in Newport News in World War I. Shortly after his arrival at the Virginia base, malaria broke out in one of the barracks. Despite his best effort, the troops refused to take quinine. At last, he hit on the idea of establishing a nighttime anopheline mosquito patrol. "After they got to bed at night," Freeborn reported, "we went around to see that they were tucked in. We had a squad of men that collected all the mosquitoes they could find in the barracks. The mosquitoes would go to the windows. The squad went through with brooms and boxes and collected all the mosquitoes. . . . After this scheme of control was in effect we didn't have a single case of malaria."[85]

Williams believed that DDT offered an inexpensive chemical means of long-term residual control. The USDA experiments in Stuttgart, Arkansas, in 1943 and 1944 demonstrated that a single application of DDT around the windows and doors of a building provided months of protection against *An. quadrimaculatus*. Shortages of DDT prevented the implementation of the residual spray program until after the war. In August 1945 the War Production Board approved the use of DDT for civilian projects. During the next two years, spray teams applied DDT treatments to millions of homes throughout the Southeast.

While Williams and Freeborn were planning the postwar malaria eradication program, Thomas Headlee, Tommy Mulhern, and Robert Glasgow were laying the groundwork for the creation of a national mosquito control organization. In 1935 Headlee organized the first meeting of the Eastern Association of Mosquito Control Workers (EAMCW). At first, the organization's primary mission was "the protection of mosquito-control work against attacks by ill-advised individuals and groups." Headlee and Mulhern, however, never lost sight of the goal of using the EAMCW to contribute to the "education of mosquito-control workers."[86]

The idea of using the EAMCW as the nucleus for a national organization to represent the mosquito crusade surfaced in 1939. Each year Tommy Mulhern organized one or two field trips for the EAMCW's members. In January 1939 Mulhern and Headlee led an eleven-day trip to Georgia and Florida. They spent their final days in Florida at the USDA Orlando research station. Tommy Mulhern reported that Bishopp, Willard King, and T.H.D. Griffitts told the EAMCW group that they were "all agreed that there seemed to be a demand for a nation-wide organization."[87]

In 1940 Lester Smith, Robert Vannote, and Ralph Vanderwerker prepared a "trial number" for a national publication. In November the members of the EAMCW approved the publication. In March 1941 the EAMCW launched a quarterly publication, *Mosquito News*. Headlee explained that *Mosquito News* would serve as the "house organ" for the Eastern Association. Its immediate objective was to provide a "service to the membership" by "keeping all informed and giving everyone a chance to describe his problems." The long-term goal was to foster a "unified effort to accomplish a common purpose."[88]

On November 30, 1942, a special committee of the EAMCW met in New York City to discuss the *Mosquito News* format. The discussion over editorial questions turned into a heated debate about the Eastern Association's future.

The participants broke into three factions. One group maintained that the EAMCW should remain unchanged; others proposed that the Eastern Association affiliate itself with the American Association of Economic Entomologists; finally, Robert Glasgow, New York's state entomologist, proposed the creation of an "American Association of Mosquito Control Workers." At the end of the meeting, the special committee voted to make no changes in *Mosquito News*.

Two weeks later, Robert Glasgow wrote a seven-page, single-spaced letter to Tommy Mulhern describing his frustration and disappointment with the special committee meeting. "I think you and I have known each other long enough to speak with absolute frankness," Glasgow declared, " . . . so I shall not be mealy-mouthed about calling a spade a spade." Glasgow reminded Mulhern that during the debate about forming a national association, "one member of our special committee protested that if we admit too many of the southern and western workers to full, active membership, we might 'lose control' of the Association."[89]

Glasgow believed that the anti-mosquito movement had reached a turning point. Historically, New Jersey had led the mosquito crusade. "I recognize fully our debt," he observed, "to the New Jersey group for its magnificent contribution in work, service, and even expense." Times had changed. The old leaders of the mosquito movement needed to put aside their "provincial loyalty."[90]

Glasgow ended his letter on a somber note. He told Mulhern that he feared that an "unwholesome" attitude of "distrust and incipient hostility" would prevent the formation of a national association. "This very frank letter," Glasgow declared in closing, "has not been easy to write; I hope it may not be misunderstood. I am writing it to you because, as secretary, you are close to our policy making officers, and because I think you will present it fairly to them and to others."[91]

During the next twelve months, Tommy Mulhern championed the idea of forming a national association. By March 1943 Mulhern won Headlee's and Vannote's support for the formation of a special committee to oversee the EAMCW expansion into a national organization. A year later, the members of the EAMCW approved the formation of the American Mosquito Control Association (AMCA). Robert Vannote was elected the AMCA's first president, and Robert Glasgow agreed to serve as editor of an expanded *Mosquito News*.

World War II produced profound changes in the anti-mosquito movement. New mosquito-borne diseases emerged to threaten the public's well being. On the home front, the MCWA succeeded in protecting military personnel and war workers from the menace of malaria and encephalitis. Overseas, medical entomologists and engineers organized the most effective wartime anti-mosquito campaign in history. By the war's end, USDA researchers had discovered powerful insecticides that promised a new era in insect control.

Headlee and Herms retired shortly before the end of the war. In many respects, these men embodied the mosquito crusade. Each man could look back at his career with satisfaction. Under Headlee's leadership, Rutgers had become a center for mosquito control research. Headlee was confident that Tommy Mulhern would continue to guide the New Jersey Mosquito Extermination Association and the newly formed American Mosquito Control Association.

Herms took pride in his role in making medical entomology into a "man-sized" profession. Thirty-five years had not dulled his missionary fervor. He looked forward to the future. "We may," he declared, "expect the white heat of war to refine this phase of entomology. . . . Entomology has contributed much to the welfare of mankind. I have seen much of its fruition during my lifetime." He was confident his students would help make the postwar world a better place "by combating the louse, the flea, and the mosquito."[92]

Chapter 8 The Postwar Era

ON FEBRUARY 7, 1949, 162 mosquito fighters assembled in the lecture auditorium of Agriculture Hall on the campus of the University of California at Berkeley. The occasion was the seventeenth annual meeting of the California Mosquito Control Association (CMCA) and the first West Coast meeting of the recently incorporated American Mosquito Control Association (AMCA). During the next three days, the topics under discussion ranged from the growth of new state programs to the role of DDT in mosquito control. Charles Hutchinson, dean of Berkeley's College of Agriculture, delivered the University's official welcome. "The University of California is justly proud of its accomplishments in mosquito control," Hutchinson declared. "Although New Jersey control work began in 1904, a year before the initial work of Prof. H. J. Quayle of this university in 1905, we have consoled ourselves with the thought that New Jersey probably needed the work more than we did."[1]

At the end of Hutchinson's opening remarks, Harold Gray, superintendent of the Alameda Mosquito Abatement district and the AMCA's incoming president, presented the university with a portrait of William Herms that "the Prof's" former students and colleagues had commissioned as a tribute to their mentor's life-long service to the mosquito crusade. The picture was iconic. The artist depicted Herms standing before a blackboard in a college lecture hall. Three books representing his work as a medical entomologist were placed in the foreground. On the blackboard, the word "Malaria" was written in bold letters along with an outline of the Penryn campaign and the life cycle of a malaria parasite.

Later that morning, Herms delivered his final address to the mosquito warriors. Harold Gray, his friend and "teammate" for nearly forty years, had invited him to provide a retrospective on a life spent fighting mosquitoes.

Figure 8.1. Portrait of William Herms presented on February 7, 1949. Reproduction courtesy of Bancroft Library, University of California, Berkeley.

Herms, who died twelve weeks later, titled his talk "Looking Back Half a Century for Guidance in Planning Mosquito Control Operations."

Herm's valedictory to the mosquito crusaders was notable both for its perspective on the history of the mosquito crusade and for Herms's injunctions for the future. Herms spoke modestly of his career. Many had contributed to defeating "the Minotaur of California." "I do not infer that mosquito control alone deserves credit," Herms observed, "for the notable reduction of malaria, but I do contend that anopheline mosquito control work had more to do with it than some folks are willing to concede."[2] As for the future, Herms admonished the mosquito warriors to "*Know well the insects*" and "*Keep faith* with the folks who look to you for the accomplishments of mosquito control" [emphasis in original].[3] The mosquito control movement must never forget these imperatives. "Yours is, therefore, a serious responsibility based on *good will*," Herms concluded. "Your constituents look expectantly and confidently to you for the abatement of annoying and disease-bearing mosquitoes. You have a job to do—do it with sincerity and devotion. Continue to be worthy of the good will of your constituents."[4]

Fueled by the availability of DDT and, later, by the subsequent discovery of other powerful chemical toxicants, the anti-mosquito movement experienced spectacular growth during the two decades following World War II. This period, the insecticide era, brought an explosion of new programs from the Gulf Coast to the Pacific Northwest. New state and regional anti-mosquito associations formed in Virginia, Utah, Texas, Louisiana, Oregon, and Washington. The greatest growth took place in California and Florida. By the beginning of the 1960s, California and Florida's anti-mosquito programs towered over the rest of the nation both in the size of their operations and their commitment to research in mosquito control.

The expansion of the mosquito control movement, however, came at a great cost. The publication of Rachel Carson's *Silent Spring* in 1962 exposed a current of discontent with mosquito control's reliance on pesticides that was to grow in the 1960s. On the first Earth Day, April 22, 1970, Herms's darkest fears were realized. Protesters marched in Chicago, Boston, San Francisco, Los Angeles, Fort Lauderdale, and hundreds of other American cities calling for a ban on DDT and stringent restrictions on mosquito control programs.

There is a tragic quality in the development of the mosquito control movement in the decades following World War II. DDT and the other insecticides that sparked the spectacular growth in the 1940s and 1950s were responsible for the public's disenchantment with mosquito control in the 1960s. The remarkable thing is that, at the beginning of the insecticide era, the leaders of the mosquito crusade cautioned against an over reliance on DDT.

Harold Gray played the part of Cassandra in his presidential address at the 1949 AMCA meeting in Berkeley. Gray, who was a civil engineer, warned against assuming that there was any single, ready-made solution to mosquito abatement. World War II and the discovery of DDT, Gray told his listeners, had opened a new stage in the mosquito crusade. In the four years since the end of the war, mosquito control had come to rely on the miraculous insecticides. "The disturbing feature of some programs [of mosquito control]," Gray declared, "is that so much effort and expense is being lavished on insecticidal operations of only temporary effect." Permanent control and source reduction work was discontinued. Insecticides, Gray cautioned,

> are toxic agents. Our experience with the new materials is relatively short in time. Most of us have been convinced by experience that if DDT is used in appropriate situations and with reasonable care it is not directly harmful to man. But when it is spread broadcast over the country, what may be its cumulative effects in time? Frankly, we don't know. We do not know the ultimate effect that accumulations of these insecticides may have upon the ecology of many of the lower orders of

> animals which are most affected by them. . . . Are we doing something to upset the ecological balance? Most of us don't know, and some don't seem to even recognize that this may be a matter of great economic importance. . . . The reckless and ill-considered broadcast of these insecticides could produce conditions which might result in extreme public disfavor, and possibly restrictive or punitive legislation, which could impede sound mosquito work for many years.[5]

It was only prudent, Gray advised, "to pause, take stock, and if possible see where we have gone and where we are going. If we, like a mariner driven by storm, have been blown off our true course, now is the time to take bearings and set our course aright again."[6] Gray's parting recommendation was simple and direct. "Let's stick to sound fundamentals. Let's do as much basic control as possible in the prevention of mosquito production. . . . With our new insecticides, we do not know, as yet, what the end results, the cumulative effects, may be. Caution and common sense are needed."[7]

Dick Peters, Tommy Mulhern, and Maurice Provost discussed Gray's speech and the future of the mosquito control movement in the days immediately following the Berkeley conference. Peters had organized a five-day postconference motorcade from Berkeley to Los Angeles to survey California's mosquito program. On February 10, Peters, Mulhern, and Provost, along with forty-nine other mosquito warriors, left the Hotel Claremont in a ten-automobile caravan. There were no assigned seats. Peters encouraged the participants to switch cars whenever they stopped and meet as many people as possible.

Provost later summarized his experiences on the motorcade in a talk before the Florida Anti-Mosquito Association (FAMA). Provost held a special fondness for the Berkeley campus. Years earlier, Provost had shared a dormitory room with Aldo Leopold's son on the UC campus. During World War II, Provost and Jack Rogers had served as John Mulrennan's point men in organizing Florida's component in the USPHS's Malaria Control in War Areas program. After the war, Provost returned to graduate school and earned his Ph.D. at the University of Iowa. In 1947 John Mulrennan persuaded Provost to return to Florida to take charge of the mosquito research effort of the Florida State Board of Health's new Division of Entomology.

Provost looked forward to the motorcade. "Hearing and reading about mosquito conditions is fine," Provost observed, "but eyes are always the best teacher." Seeing California's mosquito program firsthand would be instructive. "The main objective of the motorcade," Provost declared, "was to get all these people together with abundant opportunity to exchange ideas. And this it did admirably. You spend a whole day in a car sitting next to a man and you're bound to get well acquainted."[8]

As the caravan made its way to Woodland and Sacramento, Peters and Provost discussed California's subvention program and the Public Health Department's new Bureau of Vector Control. Later, near Fresno, Mulhern shared his impressions on the future of the control movement. Mulhern, who had begun his career in mosquito control as a high school student working as a mosquito inspector in Monmouth County, had gone on to become Thomas Headlee's right-hand man at Rutgers. In the 1930s Headlee and Mulhern had developed the New Jersey light trap. Provost, who was preparing to launch a series of surveillance experiments, had a number of questions about the light trap. The motorcade, Provost later observed, had fostered "a constant give and take of information in a congenial atmosphere. You're bound to learn something. And all on the motorcade did learn plenty."[9]

Provost was struck by the differences between the mosquito control programs in California and Florida. California's dry summers and cool winters stood in sharp contrast to Florida's wet, subtropical climate. The differences between the two states approaches to mosquito control were equally striking. California did not have "much of a natural problem" with mosquitoes. The increase in rice cultivation, irrigated crops (tomatoes, grapes, etc.), and pasturelands created vast opportunities for mosquito breeding. This expansion had taken place in the San Joaquin and Sacramento valleys. By 1949 thirty-nine of California's forty-two mosquito abatement districts (MAD) were located in the Great Central Valley. These programs relied on larviciding as their chief means of suppressing the mosquito pest.[10]

Florida's mosquito problem grew out of natural conditions. Accordingly, before World War II local mosquito control was based on ditching and drainage. The discovery of miracle pesticides during the war led to a fundamental change in mosquito control work in the Sunshine State. After DDT became available for civilian use, most of the state's ten MADs abandoned source reduction work. In the 1950s a generation of Floridians grew accustomed to thermal fog mosquito control trucks trailing clouds of DDT on their mission to kill adult mosquitoes.

Like Gray, Provost worried about the long-term environmental effects of pesticides. That was one reason he was interested in seeing firsthand what was happening in California. Provost's experiences in California left a deep impression. More money was spent on mosquito control and mosquito research in California than in any other state. By 1948 the combined state and local allocations for mosquito control in California exceeded $2 million. This was almost three times the amount spent on mosquito control in New Jersey ($750,000) and five times the total expenditures in Florida ($400,000).[11]

Two factors account for the tremendous growth in mosquito control in the early postwar period. First, many public health officials feared that returning military personnel might reintroduce mosquito-borne diseases such as malaria and dengue fever in areas where these diseases had previously been endemic. Additionally, some public health workers worried that returning GIs might carry new mosquito-borne diseases like Japanese encephalitis back to the United States. The availability of inexpensive, powerful insecticides was the second factor fostering the dramatic expansion in mosquito control. In August 1945 the War Production Board released DDT to civilian use. A year later, California MADs were annually spraying more than 100,000 pounds of DDT across the state.[12]

The USPHS was one of the catalysts for the use of DDT after the war. USPHS officials grew alarmed in 1943 over reports of outbreaks of malaria near prisoner-of-war camps and military hospitals. Mosquito control workers in Utah organized a campaign to suppress anopheline mosquitoes in Box Elder County after a sharp increase in malaria at a military hospital. In 1945 public health workers in New Jersey identified fourteen hundred cases of malaria in returning GIs. Five years earlier there had been only twenty cases in the entire state.[13]

The USPHS launched the Extended Malaria Control Program in 1945. The goal was to protect civilians from mosquito-borne diseases in areas where malaria and encephalitis had previously been endemic. Like the MCWA, the program focused on reducing the malaria threat in the thirteen southeastern states with the highest prewar malaria rates. By the end of its first year of operation one thousand federal workers had given residual DDT spray treatments to four hundred thousand houses in 110 counties in the south. On July 1, 1946, the surgeon general transferred responsibility for the extended program to the newly created Communicable Disease Center (CDC) in Atlanta. During the next twelve months, twenty-five hundred CDC and state inspectors treated more than one million houses with DDT.[14] Congress enacted legislation in 1947 that expanded the CDC's anti-mosquito mandate. The new law gave the CDC $4.6 million to eradicate malaria in five years.

The Extended Malaria Control Program and the Malaria Eradication Program brought mosquito control to much of the South. Recently, historians such as Margaret Humphreys have questioned the significance of the residual DDT treatment program in eliminating malaria. Humphreys contends that malaria was already declining in the rural South because of the migration of rural population to urban areas.[15] The effectiveness of the CDC's malaria eradication program will remain a disputed topic. What is not in doubt is that millions of individuals whose homes received DDT treatments judged the program an unqualified success. Homeowners welcomed the spray teams. The extended program "sold itself."[16] The residual treatments brought relief from mosquitoes and other insect pests. At the very least, the CDC malaria eradication program generated widespread public approval for the use of insecticides as a tool in mosquito control.

Concern over the possibility of the spread of the western equine encephalitis (WEE) and St. Louis encephalitis (SLE) viruses and the possible introduction of Japanese encephalitis led to the rapid expansion of mosquito control in California in the postwar era. In 1944 the California legislature asked the University of California, the California Mosquito Control Association (CMCA), and the Department of Public Health to prepare "A Report on Investigations of the Disease Bearing Mosquito Hazard in California." A committee consisting of Bill Reeves, K. J. Meyer, and Frank Stead, the director of the health department's Bureau of Sanitary Engineering, took charge of the study. The committee issued its report in 1945. The committee "urged that the control programs and research be broadened to include all disease vectors and that mosquito control be extended geographically." They recommended that the California legislature approve funding for a statewide subvention program for the mosquito abatement districts as well as funding for mosquito surveillance and research at the Department of Public Health.[17]

In 1946 California's lawmakers approved Assembly Bill #28 authorizing a six hundred thousand dollar biannual subvention for mosquito control. An encephalitis epidemic in 1945 with 240 documented cases and fear of the introduction of Japanese encephalitis into California had added urgency to the debate. The USPHS asked Don Rees and Bill Reeves to go to Japan and prepare a report on the disease. While Rees and Reeves were studying the problem, the California legislature passed Assembly Bill #28. This law gave the Department of Public Health responsibility for allocating the subvention funds. Assembly Bill #28 stipulated that the state money could only be used in areas where there was a demonstrated threat of a resurgence of malaria or outbreaks of encephalitis. Moreover, MADs must demonstrate that they employed either a full-time entomologist or contract entomologist in developing their control program. The legislature amended Assembly Bill #28 in 1947 and voted to provide six hundred thousand dollars annually.[18]

The state subvention program triggered a spectacular expansion in the size and scope of mosquito control in California. There were twenty-nine MADs covering forty-six hundred square miles of the state in 1945. Three years later, the total area under mosquito control had expanded to more than fifteen thousand square miles while the number of MADs had increased to forty-two. By 1954 there were fifty-three agencies covering thirty thousand square miles.[19]

Assembly Bill #28 added impetus to efforts within the Department of Public Health to consolidate oversight for mosquito and vector control in a single state agency. Dick Peters noted this when he returned to civilian life in January 1946. Peters had hoped to resume work in the health department's mosquito control program. There was, however, no opening in the health department's mosquito section, so Peters was given a position as a health inspector with the Bureau of Sanitary Engineering.

Peters discovered that there had been significant changes in the health department. During the war, Wilton Halverson, who had served as Los Angeles County health officer, had become the state health director. Halverson had brought his assistant, Frank Stead, with him to the health department's headquarters in Berkeley. Halverson placed Stead in charge of the Division of Sanitary Engineering. One of Stead's first acts was to win Halverson's approval for a reorganization plan. Stead's proposal consolidated all of the Department of Public Health's sanitation work in the newly created Division of Environmental Sanitation. In the new organizational plan, the functions of the old Bureau of Sanitary Inspection and the mosquito and vector control work of the Bureau of Sanitary Engineering were combined in a single organization. At the time, the Bureau of Sanitary Inspection was the larger of the two organizations.[20]

Assembly Bill #28 led to a dramatic expansion in mosquito and vector control work. The subvention funds allowed Stead to complete his reorganization plan. On July 1, 1947, Stead named Arve Dahl head of the Bureau of Vector Control (BVC) in the Division of Environmental Sanitation. Stead and Dahl had met when they were students at Harvard. Dahl, who was a civil engineer, lacked experience in mosquito control. Accordingly, Stead assigned Dick Peters responsibility for the BVC's biological section.[21]

Peters remembered the period immediately after the war as a time of tremendous activity. He spent 90 percent of his time providing technical assistance to the exiting mosquito abatement programs and to encouraging the formation of new control districts. Peters later observed that during this period he "almost became a stranger to his family." He was "constantly on the go" traveling from northern to southern California rallying support for mosquito control.[22]

One of Peters's first challenges came in the northern California counties of Sacramento and Yolo. He thought it essential that the counties join in a single mosquito abatement district. Sutter and Yuba counties had formed California's first two-county district in February 1946. Peters believed that Sacramento and Yolo counties should follow this precedent. Neither county could solve the mosquito problem alone. The expansion of irrigated pasture and rice fields in rural Yolo County would produce tremendous broods of mosquitoes.

Unfortunately for the more populous Sacramento County, mosquitoes were notorious for not respecting political boundaries. Peters argued that a two-county district made both entomological and economic sense.[23]

During his first six months with the Bureau of Sanitary Engineering, Peters made repeated trips to Sacramento and Yolo counties in his effort to win support for the two-county MAD. He was surprised when the commissioners asked him if he would become the district superintendent. More discussions followed. He was on the verge of accepting the post when he had an interview with one of Sacramento County's most powerful politicians. The discussion went well until the interview's conclusion. The county supervisor assured Peters that he had his support. There was, however, one condition. Peters must promise not to spend one penny of Sacramento tax revenues in Yolo County. Peters refused and withdrew his name from the search process. A few weeks later, the counties' supervisors announced that they had hired George Umberger as the MAD's superintendent. Peters, who was never prone to dissemble, made it clear that he believed Umberger had agreed to the commissioner's stipulation. It was, Peters later declared, the beginning of "what wasn't a beautiful friendship" with George Umberger.[24]

Personalities, politics, and pesticides dominated the relationship of the BVC and the California MADs. The subvention funds gave the BVC a powerful lever. Old directors grumbled at the bureau's directives. They resented the fact that Assembly Bill #28 stipulated that programs receiving subvention must employ a trained entomologist in developing their control program. Nevertheless, they wanted the state money, and twenty districts applied for subvention by the end of the program's first year.

There were numerous skirmishes. One of the first fights involved the distribution of war surplus equipment. Peters learned that Umberger had used his political connections in Sacramento to get control of "every last jeep" in the war assets program. Peters protested that other districts needed equipment. Umberger, however, refused to share the equipment. A heated exchange ensued. Umberger eventually backed down when Dahl and Peters threatened to take the matter to the governor's office.[25]

Tension between the BVC and the MADs came to a head in 1950. A handful of the most disgruntled superintendents organized a clandestine meeting, which they invited Bill Reeves to attend. Reeves, who maintained the Hooper Foundation's encephalitis research station in Bakersfield adjacent to the Kern County MAD, had excellent relations with the local programs. After the meeting, Reeves received a phone call from the meeting's organizers. They told him that the group had decided after he had left that they wanted him to go to Wilton Halverson, the director of the Department of Public Health, and state their position. Arve Dahl was running the BVC as if he was the "czar for all the mosquito control in California." The district managers declared that they would no longer tolerate Dahl's imperious ways.[26]

Reeves agreed to serve as their emissary. He arranged a meeting with Halverson. Halverson asked Frank Stead and Arve Dahl to attend. Stead and Dahl listened to the charges. When Reeves finished reciting the mosquito managers' complaints, Dahl launched his rebuttal. Halverson interrupted, "Mr. Dahl, we're going to change our ways because the State Department of Public Health shouldn't be having these relationships."[27] Six months later Dahl resigned from the BVC, ostensibly to accept the position as the state civil defense director. Dahl preferred dealing with bomb shelters than irate mosquito control managers. Stead named Dick Peters the BVC's new head. Peters remained in this position for the next twenty-seven years.

Peters's vision of the BVC was radically different from Dahl's. Dahl was an engineer who believed that every problem had a technical solution. When the managers balked, he used the threat of withdrawing subvention funds to force their agreement. Peters knew that he could not force the districts to embrace sound entomological practices. If William Herms had taught him anything, it was that you have to educate people rather than try to beat them into submission.

Peters had begun his effort to woo the managers even before he became the BVC's director. In 1948 the BVC began publication of a monthly newsletter called *The Buzz*. The newsletter was Peters's effort to reach out to the districts. He dictated the early issues to his secretary who typed them directly onto a stencil and mailed copies. Peters made sure that every issue contained a feature highlighting the history and accomplishments of one of the MADs.

The Buzz was more than an informal newsletter. Peters used it to raise serious issues. By 1948 California's MADs were annually spraying over hundred thousand pounds of DDT. Many of the districts had abandoned source reduction and permanent control work. Ted Raley, the first manager of the Sutter Yuba District, modified a war surplus jeep into what became known as "the plumber's nightmare." The jeep's exhaust system became a DDT spray system. Most of the districts employed versions of Raley's "plumber's nightmare" in their larviciding and adulticiding work.[28]

Peters opposed the districts' growing dependence on chemical control. He agreed with Provost's assessment of the differences between California and Florida. Nature produced the mosquito problem in Florida; human beings were responsible for the problem in California. Poor planning coupled with the growth in irrigated pasture and the cultivation of rice, cotton, and other crops produced tremendous new sources of mosquitoes. Chemical control,

Peters contended, offered at best a temporary solution. At worst, he feared that overreliance on chemical control would undermine the statewide anti-mosquito movement.

Like Provost, Peters believed that research was essential. In the weeks following the 1949 Berkeley meeting, the BVC organized an operational research program to explore new ways of mosquito control. None of the districts had the resources needed for such a research program. Ed Washburn, manager of the Turlock abatement district and a former USPHS health officer, volunteered use of facilities at the Turlock headquarters for Peters's first research station. In 1949 Peters recruited a young entomologist named Earl Mortenson to guide the research work at Turlock.

Mortenson was part of a new generation of mosquito workers. After the war, Mortenson had used the GI Bill to earn an undergraduate degree in entomology from Colorado State University. In 1948 he started graduate work at Berkeley in medical entomology and parasitology. Mortenson's advisor told him that he should not waste time earning a master's degree. He should set his sights on a doctorate. "I found out later on, as I proceeded along the way," Mortenson explained, "that it would take me a good three or four years to do that. By that time, I was running out of money and I needed to get out in the real world. I decided to start looking for a job. I heard about the Bureau of Vector Control, which is right across the street from us."[29]

Mortenson earned a reputation for his helpfulness in the mosquito districts. On one occasion Mortenson received a call from the Fresno health office requesting his assistance in solving a politically delicate mosquito problem. A homeowner had filed numerous complaints about mosquitoes. The Fresno sanitarian asked Mortenson if he would drive out to the house and see if anything could be done. Mortenson discovered that the house was, in fact, a bordello on "the wrong side of the tracks." It was the height of the cantaloupe season and the county was filled with farm workers. The brothel was doing a booming business until the mosquitoes arrived.[30]

Mortenson inspected the bordello and identified the source of the mosquitoes in the run-off from ten "swamp coolers" which provided limited air-conditioning for the house. The discharge water from the coolers formed a standing pool beneath the house making an ideal habitat for *Culex* mosquitoes. At the end of his visit, Mortenson explained the problem to the house's madam. He told her that she should hire a plumber to correct the problem. "That's a good idea," the madam declared. "I'm going to get a hold of that plumber. He owes me a lot of favors!" Before leaving, Mortenson crawled under the house and sprayed the water with insecticide. He warned that this was only a temporary solution. Unless the water was removed, the mosquitoes would return. A few weeks later, the madam phoned the mosquito district's headquarters gushing with praise for the young man who had solved her mosquito problem. Business had never been better. Both her clients and the "gals" were enjoying mosquito-free trysts. Mortenson later told Peters that this was clear evidence that effective mosquito control encouraged economic activity.[31]

Not all of Mortenson's forays into economic entomology were as successful as the Fresno bordello episode. Shortly after Mortenson joined the BVC, Peters asked him to perform a series of flight range studies on *Aedes nigromaculis* mosquitoes. Mortenson sprayed a mixture of reddish-orange Rodimun B florescent dye on an irrigated pasture near Modesto and "scattered" thirty New Jersey light traps throughout the county. The next morning Mortenson received quite a surprise when he returned to pasture and found a herd of pink-nosed dairy cows standing in the treated pasture. During the night, the cows in a neighboring field had broken through the fence. Mortenson remembers having to go to the farmer and "talk to him about it. It turned out that the mosquito abatement district bought some pink milk for awhile." The experiment, however, worked. Mortenson reported to Peters that *Ae. nigromaculis* mosquitoes had a flight range of four or five miles.[32]

Earl Mortenson was not the only new member of the BVC in 1949. During the motorcade after the Berkeley meeting, Peters learned that Mulhern wanted to leave New Jersey. Mulhern's discontent dated from Headlee's retirement in 1943. Mulhern had been Headlee's right-hand man. Initially, Bailey Pepper, who succeeded Headlee as the entomology department's chair, allowed Mulhern to guide the day-to-day mosquito control work at the Station. He needed Mulhern's contacts with the mosquito commissions and the New Jersey Associated Executives in Mosquito Control. It was clear, however, that Pepper intended to change things at the experiment station.

Bailey Pepper had little experience in mosquito control. Trained as an agricultural entomologist, Pepper graduated from Clemson and earned his master's degree at Ohio State before completing his doctorate at Rutgers in 1934. Pepper, who retained his South Carolina accent and Southern manners throughout his forty-year career at Rutgers, resented Mulhern's status as the de facto head of the experiment station's mosquito program. Pepper was powerless as long as Headlee was in charge.[33]

Pepper had reason to fear Mulhern; his influence in the mosquito movement stretched across the continent. Beginning in 1935 with the formation of the Eastern Association of Mosquito Control Workers (EAMCW), Mulhern had rallied support for a national mosquito organization. His position as the EAMCW's secretary gave him an opportunity to encourage the formation of new state and local programs. Nine years later Mulhern was one of the prime movers in the creation of the American Mosquito Control Association (AMCA). Pepper worried that Mulhern's national standing would undermine his standing at Rutgers.

Mulhern's frequent trips to other states to encourage the formation of new programs added to Pepper's concern. As Mulhern had anticipated, the end of the war sparked the formation of new mosquito control organizations across the nation. For example, mosquito control in Virginia had languished after World War I. In 1933 Louis Williams had recruited Roland Dorer from New Jersey to serve as the assistant state director of the CWA's malaria control effort. Dorer, a Rutgers graduate, found an early ally in Norfolk in a minor league baseball club owner named Perry Ruth. Dorer and Ruth set about organizing a mosquito control program along the Chesapeake. In 1940 the Virginia legislature approved a mosquito abatement act patterned after the New Jersey law. The first district to form under the new law was in Virginia Beach and Princess County. Additional programs started in Elizabeth County (1945), Warwick County (1946), Newport News (1946), and Norfolk (1947). Dorer and Perry Ruth organized the Virginia Mosquito Control Association (VMCA) in 1947. Dorer and Ruth's first victory came later that year when legislators in Richmond authorized $56,000 subvention for mosquito control.

The tension between Pepper and Mulhern intensified in the late 1940s. In 1948 Pepper announced his decision to reorganize Rutgers's mosquito control unit during an impromptu speech at the New Jersey Mosquito Extermination Association's annual meeting. "I realize," Pepper declared, "that some of the members in this organization wonder as to what my stand is and where I come into mosquito work, and with that in mind I would like just a moment to outline very briefly the setup in our department." There had, Pepper acknowledged, been "some discord as you people know—differences of opinion" when it became known that Pepper planned to reorganize the mosquito control program. He acknowledged that he lacked a "technical background in mosquito work."[34]

Pepper was confident in his ability to lead the state's mosquito control program. He announced that he had asked Daniel Manley Jobbins to "serve as coordinator . . . of all phases of mosquito work." Jobbins, who possessed a talent for getting along with people, had earned his master's degree at Rutgers in 1935. While a graduate student, Jobbins had assisted Frank Miller in organizing the experiment station's contacts with the commissions and mosquito research. Jobbins continued this work until 1938 when he accepted a position as a medical entomologist at the Gorgas Memorial Laboratory in Panama. Jobbins served during World War II as a sanitarian for the USPHS and as a medical entomologist for the Caribbean Section of the Pan American Sanitary

Bureau.[35] In 1948 Pepper passed over Mulhern and offered Jobbins the position leading Rutgers's mosquito control work. Pepper was aware that his decision would not please everyone. "I wish that everybody in the department," he declared in a thinly veiled reference to Mulhern "could be boss." That, however, was not possible.[36]

Mulhern resented his demotion. He had devoted nearly twenty years to his work at the station. The conflict between the two men was probably inevitable. When Pepper became New Jersey's state entomologist, Mulhern was known as "Mr. Mosquito Control." Both men had large egos. Mulhern's close relationship with the superintendents and mosquito commissioners forced Pepper into a subordinate role. Hiring Jobbins was Pepper's way of demonstrating that he was in charge.

Mulhern began making plans to leave New Jersey after Pepper's announcement. In 1949 the Texas legislature passed a law authorizing the formation of mosquito abatement districts in counties along the Gulf Coast. Officials in Jefferson County asked for Mulhern's advice on building an effective program. Mulhern visited Jefferson County and prepared a report. He declined the county commissioners' offer that he take charge of the new abatement district.

The 1949 Berkeley meeting convinced Mulhern that he should leave New Jersey. California and Florida were poised for tremendous growth. Mulhern stopped in Utah on his way home from the 1949 Berkeley meeting to attend the Utah association's second meeting. Mulhern and Harold Gray, the AMCA's new president, applauded Don Rees's efforts to build a statewide movement. Rees had supported Mulhern's efforts transform the EAMCW into a national organization in 1944. Later Rees helped draft the AMCA's constitution, served as a member of the new association's executive board, and led the association's Interim National Board. The previous year Rees provided the impetus for the formation of the Utah Mosquito Abatement Association (UMAA).[37]

On the flight from Salt Lake back to New Jersey, Mulhern weighed his options. His relationship with Pepper was tenuous. Mulhern felt that Pepper resented his prominence in the mosquito control movement. Before Mulhern left California, Dick Peters had asked him if he would be interested in leaving New Jersey. Peters told him that he was looking for someone to lead the BVC's mosquito work in the Central Valley. Mulhern's conversations with Don Rees and Harold Gray at the Utah meeting convinced him that in the next important chapter in the mosquito crusade would take place on the West Coast. By the time Mulhern had reached New Brunswick, he had decided to leave Rutgers and accept Peters's job offer at the BVC.

Mulhern's move to California marked a transition in the history of the mosquito crusade. During the first half of the twentieth century, New Jersey

Figure 8.2. Tommy Mulhern holding a copy of *Mosquito News*, circa 1950. Photo by Thomas D. Mulhern provided by the Mulhern family.

had served as the anti-mosquito movement's focal point. In 1949 California and Florida emerged as the new powers within the movement. Mulhern's move added impetus to this. When Mulhern arrived in Fresno, the institutional base of the AMCA shifted to the West Coast. The headquarters of the AMCA remained in California for the next three decades.

Florida was the outstanding example of the growth of the mosquito control movement in the postwar era on the East Coast. There were only ten mosquito control districts in the state in 1945; nine years later forty-two local agencies conducted mosquito control work in Florida. For the Sunshine State, 1949 marked a watershed for the mosquito control movement. The Florida State health board celebrated its Diamond Jubilee. Sixty years earlier the health board had come into existence in the wake of the 1888 Jacksonville yellow fever epidemic. After World War II the USPHS's Extended Malaria Control Program and the CDC's Malaria Eradication Program had succeeded in eliminating indigenous malaria in Florida. Victory in the war against malaria, however, meant that it was only a matter of time before the CDC ended funding for the residual DDT spray program. The announcement of the funding cut came in 1949. Federal support for mosquito control would end on June 30, 1950.

There were other disturbing developments for the mosquito control movement in Florida and across the nation. In 1947 Fort Lauderdale mosquito control workers reported that the normal application of a 5 percent DDT solution had no discernible effect on salt marsh mosquitoes. J. H. Bertholf, director of mosquito control in Broward County, told his colleagues in the Florida Anti-Mosquito Association in 1949 that he had "decided to waste no more money on DDT."[38] The miraculous "magic dust" had lost its efficacy against the hordes of salt marsh mosquitoes along Florida's east coast. Willard King, who had returned to Florida after the war to resume his leadership of the USDA Orlando laboratory, dispatched Chris Deonier to Brevard County to study the problem. Deonier set up a field station in Cocoa Beach in the same area where he and the other entomologists at the USDA Orlando Lab had first performed their tests using DDT as a larvicide and adulticide. "It soon became apparent to Dr. Deonier," Cain reported, "that [the mosquitoes] actually were resistant to DDT."[39]

The appearance of DDT-resistant mosquitoes sent a shock wave through organized mosquito control. DDT possessed two great advantages over other control measures. It was inexpensive and it was persistent. Most of Florida's mosquito control districts continued to use DDT in the 1950s. The growing problem of DDT resistant mosquitoes, however, forced mosquito control to look for other insecticides and encouraged the development of nonchemical alternatives to mosquito control.

USDA officials in Washington proposed closing the Orlando Laboratory when DDT resistance problem emerged. Mulrennan and the leaders of the Florida Anti-Mosquito Association argued that it was essential that USDA maintain the Orlando program. They succeeded in reversing the decision.[40] By 1949 John Mulrennan had concluded that the Division of Entomology could not count on any other organization for basic research on Florida's mosquitoes. The Division must build its own research program on mosquito biology and control strategies.

The anticipated cutback in CDC funding for the malaria eradication effort, the growing resistance problem, and the possible closure of the Orlando laboratory weighed heavily on Mulrennan and Provost as they flew to California to attend the 1949 Berkeley meeting. On the trip west, Mulrennan and Provost discussed the future of the mosquito control movement in the Sunshine State.

Mulrennan and Provost believed that California's subvention program offered a model for Florida. The threatened closure of the USDA laboratory reinforced Mulrennan's conviction that Florida's Bureau of Entomology must develop its own mosquito research program. When Mulrennan returned to

Florida from California, he prepared the draft for a bill giving the Bureau of Entomology $350,000 a year to distribute to local mosquito control programs for the purchase of insecticides. In 1949 the Florida legislature and the governor signed into law what became known as State 1 Funds for mosquito control.

Mulrennan gambled that the subvention funds would spark the formation of new mosquito control districts. He believed that after politicians saw that mosquito control produced tangible results, he could secure funding for permanent or source reduction work. Mulrennan's plan, however, contained a potentially fatal flaw. State 1 Funds provided money solely for the purchase of DDT at a time when resistance to hydrocarbon insecticides was a growing problem.

Provost, who was trained as an ornithologist, warned Mulrennan that he was pursuing a dangerous course. "We have reached a point," Provost observed, " . . . where all this wonderful technology of ours is being used most inefficiently if not downright wastefully. This is especially serious here in Florida where the new and popular techniques for control of mosquitoes and sandflies are intimately related to behavior aspects of these insects about which we are totally ignorant." Provost feared that State 1 Funds might serve only to worsen the situation. "If insect control is running into a stone wall," he added, "it is a wall of biological ignorance, not a wall of insect resistance." Effective mosquito control must be based on entomological research. It was imperative that Florida follow California's lead and launch its own program of mosquito research.[41]

Mark Boyd, the renowned Rockefeller Foundation malariologist, supported Provost's position. Boyd had established the Rockefeller Foundation's Malaria Research Station in Tallahassee in the 1930s. In fact, Boyd had given John Mulrennan his first job in entomology. Boyd joined the board of Directors for Florida's Health Board when he retired. Boyd recommended that Mulrennan include a provision for state support of research on mosquito biology in a new legislative proposal.[42] The legislature approved Mulrennan's proposal (State 2 Funds) for research and source reduction in 1953.

Provost had, in fact, already begun a search for a suitable location for the entomological laboratory before the governor gave his approval for the quarter of a million dollar mosquito research center. Placing the laboratory in a centrally located site was a high priority. Because much of the laboratory's work was to be directed at salt marsh mosquitoes, it made sense to place the new facility somewhere along the east central Florida coast. Brevard, Indian River, and St. Lucie counties were likely candidates. Each had its advantages. Provost's years of work in the MCWA program had made him thoroughly familiar with this part of Florida. After a final "reconnoitering trip," he concluded that a mainland site somewhere between Sebastian and Fort Pierce would offer an ideal location for the laboratory.[43]

Provost wondered "whether in our hurry to reduce annoyance from biting insects we haven't overplayed our knowledge of those insects."[44] He argued that the cost and consequences of any proposed control strategy must be analyzed before its adoption. Provost hoped that the Entomological Research Center (ERC) would provide a basis for understanding the mosquito biology in an environmental framework. "From a biological standpoint," Provost maintained, "insect control is purposeful changing of the environment to eliminate the insect." At its best, mosquito control was "applied ecology."[45]

Provost and Mulhern regularly traded claims about whether Florida or California had the greater mosquito problem. Mulhern thought he had won the contest in the summer of 1950. Known for always having a camera with him, Mulhern had snapped a picture of Betinna Rosey in front of a heap of 345,000 mosquitoes that had been caught in a New Jersey light trap on a balmy central California night.[46] Provost sent word in September that a light trap on Sanibel Island had captured 365,696 mosquitoes. "To catch over a third of a million mosquitoes," Provost observed, "in someone's backyard in one night became more understandable when we measured egg densities in Sanibel Slough as high as 45,000 [eggs] per square foot which projected, would be two billion an acre." Provost maintained that Florida led the nation in blood-sucking pests.[47]

In addition to providing a basis for mosquito research, the State 2 Funds encouraged mosquito control districts to begin source reduction work. Districts received seventy-five cents for every dollar spent on permanent control or source reduction. Brevard County began work on the state's first mosquito impoundment in 1954. The idea grew out of John B. Smith's observations of New Jersey salt marshes. Smith had noted that not all of the salt marshes produce mosquitoes. The marshes that were covered by the daily tidal flow produced few mosquitoes. Areas above the tide line, the high marshes, were prolific breeders. Smith used ditches to drain excess water from the high marshes and to provide predacious fish access to mosquito larvae. Along east-central Florida's Atlantic coast, ditching had proved costly and ineffective if not maintained. There was, however, another way to deny the *Aedes* mosquitoes of their nurseries. If the areas where the mosquitoes deposited their eggs were flooded, the mosquitoes would be unable to oviposit. This would lead to a substantial reduction in the number of salt marsh mosquitoes without the use of insecticides.

Provost's research was critical to the source reduction work in two different ways. First, little was known about the life history of *Ae. taeniorhynchus* or *Ae. sollicitans*. Basic questions about where and how they mated, their flight range, and when they took their first blood meal remained unanswered. Second, under Provost's leadership the ERC designed a series of research projects that monitored the impoundments' environmental effect. This allowed Provost to recommend changes in the impoundment program that reduced mosquito control's impact on nontarget species.

While Provost and Mulrennan were rallying support for the ERC, Dick Peters continued his efforts to improve the BVC's relationship with the local MADs and build an operational research program. One of the critical features of the Assembly Bill #28 was the stipulation that districts must employ an entomologist to oversee their control program. The rapid expansion in the number of districts made it difficult to find competent individuals to fill the posts. Peters was always on the lookout for young entomologists willing to come to California.

In January 1951, at the California Mosquito Control Association's annual meeting, Don Rees told Peters that one of his graduate students was looking for a job. This information came at an opportune moment. Roland Henderson, the manager of the Tulare MAD, needed to hire an entomologist to continue to receive BVC funds. Rees returned to Salt Lake City and told Glenn Collett about the job. "I know you haven't decided what you want to do," Collett remembers Rees saying, "but the Tulare District is looking for an entomologist."[48]

Glen Collett, who was to lead the Salt Lake City Mosquito District for thirty-two years and later served as the AMCA's president, became involved with mosquitoes by accident. After war ended, Collett resumed his undergraduate studies at the University of Utah. He took several entomology courses, but he was unsure what he wanted to declare as his major. Don Rees offered him a job preparing the specimens of mosquitoes that Rees and Reeves had collected on their trip to Japan to study Japanese encephalitis. Collett found himself sharing an office with a young professor named Lewis Nielsen. When they finished pinning Rees's specimens, Collett had found his life's work.

Collett's first assignment in Tulare was to continue Earl Mortenson's flight range study of *Ae. nigromaculis* mosquitoes. To avoid another "pink milk" problem, Collett used radioactive isotopes to tag the mosquito larvae. When the adults emerged, Collett used Jersey light traps to collect specimens throughout the district.[49] Collett's work was not limited to his research efforts in Tulare. When nearby Hanford organized a MAD, Dick Peters and Mulhern asked Collett to oversee the entomological work. Collett's stay in California came to an abrupt end in December 1951 when he was called to active duty and sent to Korea.

While Collett was leading a ten-man entomological unit doing surveys of ecoparasites on the Korean peninsula, California experienced a statewide encephalitis epidemic. The first signs that there might be a serious mosquito

Figure 8.3. Don Rees, Archie Hess, and Glen Collett, mid 1950s. Photo courtesy of Glenn Collett.

problem had come the previous winter. Heavy snowfalls occurred in the winter of 1951–1952. Skiers were pleased. Bill Reeves worried that when the snow pack melted, the runoff might produce ideal conditions for mosquito production. Record-breaking spring rains added to his concern. By April, Reeves noted, "the Central Valley was literally underwater. The rivers all went over their banks, over the tops of levies, all up and down the Central Valley."[50]

Dick Peters received reports in the late spring and early summer that there were "an abnormally large number of cases of infectious encephalitis" in the Central Valley. By June, the majority of the twenty-one MADs in the Sacramento and San Joaquin valleys informed the BVC that they had launched emergency control measures. "The breadth of the breeding area," the Department of Public Health concluded, "exceeded the physical resources to stem a large part of the emerging adults." Three additional biological factors worsened the situation. Local mosquito control workers reported an increase in the flight range and feeding preferences of *Cx. tarsalis* mosquitoes. Entomologists had believed that the flight range of a *Cx. tarsalis* mosquito was approximately two miles. In 1952 they discovered many of these mosquitoes traveled five to ten miles from their breeding sites. Simultaneously, there was a shift in the mosquito's feeding pattern. Hitherto, this mosquito had shown a preference for "the blood of either domestic or wild fowl." During the summer of 1952 numerous reports cited *Cx. tarsalis* mosquitoes "invading homes and biting man in a manner not previously observed."[51]

DDT failed to control the mosquitoes. There were several reasons for this. First, there were simply too many mosquitoes and not enough equipment. It would have required a fleet of airplanes to spray every potential larval habitat for *Cx. tarsalis* mosquitoes in the Central Valley. More important, *Cx. tarsalis* was resistant to DDT.[52] Thus, even if there had been enough airplanes, the insecticide would likely not have controlled the mosquitoes. By midsummer the Central Valley MADs were "valiantly out spraying DDT anyplace that there was a control agency." The mosquitoes were unimpressed.[53]

On July 14, 1952, the BVC's advisory committee held an emergency meeting with the mosquito districts' managers. The participants agreed that "an epidemic of encephalitis was at hand." Two weeks later the BVC invited health officers in counties where encephalitis was endemic to a second emergency meeting. At that meeting, epidemiologists predicted that there would probably be five hundred to seven hundred cases of encephalitis. Peters announced that the BVC was mounting an intensified mosquito suppression effort throughout the Central Valley. Eight days later Gov. Earl Warren ordered that $250,000 in emergency funding be given to "Operation *Culex tarsalis*."

The 1952 California encephalitis epidemic marked a watershed in the development of mosquito control on the West Coast. By the time that the epidemic had subsided in the late fall, epidemiologists reported 729 encephalitis cases with 51 deaths in thirty-seven of California's fifty-eight counties.[54] The epidemic revealed both the strengths and the weaknesses of California's postwar mosquito control effort. The BVC demonstrated its effectiveness in coordinating the response to a statewide medical emergency. Conversely, the 1952 outbreak exposed weaknesses within the local districts and in the state health department's response to the crisis. Three valuable lessons were learned. First, the emergency control program based on larviciding and adulticiding had failed. New control strategies were needed. Second, more effort should be given to anticipating epidemics. Finally, a comprehensive floodwater plan was needed. During the next twenty-five years mosquito control districts supported efforts to construct a string of flood control dams throughout the San Joaquin and Sacramento valleys.

The 1952 California encephalitis epidemic provoked a political backlash. Assembly members and senators in Sacramento called for the formation of a special committee to investigate the BVC's actions. A number of politicians openly wondered how it was possible for the state to have spent more than three million dollars since 1946 to control mosquitoes and still have such an outbreak. Was the money wasted? Were the BVC and the local districts guilty of mismanagement of public funds?

The special legislative committee asked Bill Reeves to testify. Always a teacher, Reeves attempted to provide the committee's members with a brief introduction to entomology and epidemiology. He explained that wild birds carried the virus all over the state and that these birds formed virus reservoirs on which *Cx. tarsalis* mosquitoes fed. At that point a committee member interrupted Reeves and declared, "I'm going to pass a new law. I'm going to make it against the law for birds to move from southern California to northern California." Reeves was stunned. "It wasn't going to be very easy," Reeves gently interjected, "to get the birds to obey the law." Reeves explained that birds seem "to fly where they wanted to when it was the right time to do it."[55] Reeves suggested that a better solution might be spending more money on research.

Three years later the head of California's public health department asked the CDC to prepare a program review of the BVC's administrative and technical effectiveness. The authors of the eighty-seven-page confidential report built their assessment of the BVC on a comparison of mosquito control research and operations in California and Florida. The report praised the BVC and Florida's Bureau of Entomology in providing leadership to local agencies along "technico-educational-persuasive-stimulational lines."[56]

The CDC report underscored the rapid expansion of mosquito control in California and Florida. Both states had undergone tremendous growth in the decade following the war. The pace of growth in Florida, however, had eclipsed California by the mid 1950s. California's population was nearly three-and-one-half times the population of Florida, and the state's mosquito program covered nearly three times as much territory. State funding for mosquito control, however, had remained fixed at $600,000 per year. By 1956 Florida spent more in both absolute and relative terms than California on mosquito control. State subvention for Florida's forty-two organized mosquito control programs provided an average of 75¢ from the state for every dollar the local districts spent on permanent control. In the fiscal year 1955–1956 the Florida legislature appropriated $1,250,000 to mosquito control with local districts contributing another $500,000.[57]

Mosquito control was big business in both California and Florida. Each state's annual expenditure on mosquito control was growing. By the mid-1950s both philosophical and operational differences had emerged. In California, Reeves and Peters argued that the primary justification for mosquito control was to eliminate disease-bearing mosquitoes. After World War II and the success of the malaria eradication program, the anti-mosquito movement in Florida argued that the elimination of pest and nuisance mosquitoes was essential for the state's economic growth. In 1953 Mulrennan's argument for the multimillion dollar infusion of state funds for source reduction was strictly economic.

During his campaign to win state support, Mulrennan showed legislators a bar graph in which the number of female *Ae. taeniorhynchus* caught in light traps was placed next to the total yearly expenditures made by tourists. The graph showed that as the number of mosquitoes declined the amount of money spent by tourists increased. "Tourism," Maurice Provost succinctly observed, "is a seven hundred million dollar industry in this state—and tourists do not come here to be bitten by mosquitoes."[58] Forty years later, Mulrennan's critics objected that the reasoning behind the bar graph was fallacious. "It might be argued," an observer noted, "that a decline in the number of rattle snakes could be plotted against increasing tourist dollars."[59] They missed the point. Mulrennan succeeded in winning the subvention funds because legislators in Tallahassee genuinely believed that the control of pest mosquitoes was essential for Florida's economic progress.

The CDC program reviewers noted there were troubling signs for mosquito control in both states. There was no guarantee that Sacramento and Tallahassee would continue subvention indefinitely. In California the motivation for state support grew out of the fear that "many persons who had been infected and are carriers of mosquito-borne diseases" might return from overseas and threaten public health. The threat was real in 1945. Concern had dwindled following the war. Peters and his colleagues in the BVC faced a dilemma. Support for mosquito control depended on the public's belief that there was a threat of mosquito-borne disease. The success of mosquito control in preventing disease outbreaks in the late 1940s lessened the public's concern with mosquitoes. The mosquito crusaders' challenge was maintaining a sense of urgency in the public.

The CDC program review praised the BVC for expanding its research effort after the 1952 encephalitis epidemic. The reviewers reported that by 1955 there was "unanimity" among the district managers that the subvention had "been extremely beneficial." The BVC had established itself as a "force" offering direction and guidance for the statewide movement. The district managers noted the importance of BVC's operational research initiative, its "development of standards" for mosquito control and willingness to provide "technical services" to local programs. The report concluded that the BVC had fostered a "strong awareness of mutual interdependence" and recognition that "ineffective mosquito control in one district is . . . harmful to the interests of all districts."[60]

The most striking feature of the CDC report was the reviewers' speculation about the future of the mosquito control movement. "We consider," they concluded, "or at least hope, that over a period of time, the mosquito problem will have been reduced to more containable proportions in many counties through the accumulative effect of source control measures, and that the gradual absorption of mosquito control into health departments will be in the public interest and should be done."[61]

The CDC program review validated Peters's position. To Peters, who came to mosquito control from a background in medical entomology and public health, the primary objective of mosquito control was suppression of disease-bearing mosquitoes. The control of pest and nuisance mosquitoes was at best a way to win public support for mosquito control and at worst a distraction from fighting disease-bearing mosquitoes.

While the CDC reviewers were preparing their report, the New Jersey legislature was locked in a debate about the future of mosquito control in the Garden State. The year 1955 was a bad one for mosquito control along the Atlantic Coast. Much of the Northeast suffered from a drought in June and July. This changed when two hurricanes swept across North Carolina and made their way across the mid-Atlantic state. First the tail end of Hurricane Connie buffeted New Jersey with tropical storm winds and nearly a foot of rain. Five days later, on August 19, Hurricane Donna added nine inches more. There were howls of protest when ravenous broods of salt marsh mosquitoes descended on the storm weary Garden State.

Nine days after Hurricane Donna's departure, Sen. Frank Farley from Atlantic County introduced Senate Concurrent Resolution No. 2 calling for the appointment of a Mosquito Control Study Commission (MCSC). Jesse Leslie, the executive secretary for the Bergen County Mosquito Extermination Commission explained, "The ordinary citizen was pretty burned up over the way the mosquito boys had failed to protect him" after the storms.[62] Senator Farley made it clear that he was "not antagonistic" to mosquito control. Nevertheless, since 1902 the state had given the commissions a substantial amount of money. "Mosquito control in New Jersey," the senator reportedly declared, "was big business, real money was being spent and the people who were spending it wanted to know if they were getting full value for their dollars."[63]

The MCSC delivered its report on December 1, 1955. The committee members concluded that New Jersey's historic approach to mosquito control was "fundamentally sound and should be continued." There were critics of the study group's conclusion that the mosquito control commissions be retained as the "core of mosquito control activities." Some outspoken individuals wanted to drop the commission system in favor of a system resembling California's approach. The committee opposed such a radical change. Instead they called for "certain changes to strengthen and improve the work." They recommended that the legislature provide for "greater central control, more and better research, and greater power of enforcement."[64]

The call for increased funding for research at the agricultural experiment station had wide support. New Jersey had once led the nation in mosquito research. Jesse Leslie spoke for the majority of the district managers when he declared the current level of state support "isn't enough. . . . Take a look at the mosquito research laboratory in Vero Beach, Florida," Leslie said, "a state that went into the mosquito control business years after New Jersey, and, then, look at our own limited and cramped research facilities in New Brunswick. . . . There should be more up-to-date equipment and personnel. This is a must!"[65]

Six months after the study commission issued its report, the New Jersey legislature passed Senate Bill #14 authorizing the formation of a State Mosquito Control Commission to coordinate the state's support for mosquito control. One of the questions that had to be resolved was where to place the commission. Once again, some argued that New Jersey should follow California's lead and place oversight for mosquito control in the State Department of Health. John Zemlansky, the state health officer, testified that this was not necessary. The agricultural experiment station, Zemlansky declared, "has done yeoman service in mosquito control these many years and is capable of continuing to do so."[66] The Senate and Assembly concurred and placed the commission under the Department of Conservation and Economic Development.

The study group's other recommendations did not fare as well. Efforts to strengthen the mosquito law's enforcement and inspection features, reform the procedures for selecting mosquito commissioners, and tighten the budget approval process were defeated. Most notably, legislators slashed nearly 90 percent of the proposed increase in state support for research and control work. The original proposal had called for a six hundred thousand dollar appropriation. The agricultural experiment station would receive seventy-five thousand dollars for research and an additional twenty-five thousand dollars for mosquito surveys. The remaining five hundred thousand dollars would go to county commissions undertaking source reduction work. The appropriations committee slashed all funding for research and surveys from the budget and allowed only seventy-five thousand dollars for source reduction work. The committee added ten thousand dollars to the state air-spray program. This program was created in 1949 to help southern counties control salt marsh mosquitoes. The 1956 budget called for sixty thousand dollars for this program.[67]

The formation of the Mosquito Control Commission marked the beginning of a significant change in the conduct of mosquito control in the Garden

State. The 1912 abatement act gave the agricultural experiment station oversight of the state's mosquito commissions. Senate Bill #14 introduced a new voice into the conversation. The decision to place the commission under the Department of Conservation and Economic Development was worrisome. Given the department's regulatory functions, some feared that it was only a matter of time before the commission turned into some kind of "gestapo." Jesse Leslie claimed that the presence of the experiment station's director as an ex officio member of the commission and Bailey Pepper's role as the group's secretary ensured that the commission would assist existing mosquito control programs.[68]

Three years later Pepper, the county commissions, and the state mosquito control commission faced a major challenge with the state's first eastern equine encephalitis epidemic. In 1933 Malcolm Merrill and Carl Ten Broeck had isolated the virus in Massachusetts. A year later veterinarians diagnosed 201 cases of equine encephalitis in New Jersey. Mosquitoes infected more than 184,000 horses with the disease in 1938. Epidemiologists reported that 90 percent of the horses had died. A year later twenty-five human deaths from encephalitis in Massachusetts indicated a 65 percent mortality rate for human infections.[69]

A 1959 outbreak of encephalitis in New Jersey caused panic. Wayne Crans, who served as an entomology professor at Rutgers for forty-five years, retains a vivid memory of public hysteria. Drivers avoided the Garden State Parkway because of fear of mosquitoes in the Pine Barrens. State police carried insect repellent in their cars to spray on motorists who had left home unprepared. Radio stations advised commuters "to tape quarters for your toll to the outside of your window so the toll collector could get them without opening your window." There were, Crans recalled, rumors of individuals dropping dead on the Boardwalk in Atlantic City at the height of the outbreak in September. Hotels and restaurants along the Jersey Coast closed because of the decline in tourists.[70]

The 1959 New Jersey encephalitis episode exposed new aspects of the intimate connection between wild birds, mosquitoes, and human beings that Reeves and Hammon had discovered in 1940 in the Yakima Valley. Reeves and Hammon's work on the WEE virus and *Cx. tarsalis* encouraged other researchers to explore the role of other pest and nuisance mosquitoes in transmitting viruses such as eastern equine encephalitis and St. Louis encephalitis. In 1949 Howitt found the eastern equine encephalitis (EEE) virus in specimens of *Mansonia perturbans* mosquitoes collected in Georgia. A year later, Chamberlain isolated the virus in *Culiseta melanura* in wild specimens in Louisiana.[71]

The public's fear of future arbovirus outbreaks led the New Jersey legislature to increase its support for mosquito control.[72] In January 1960 the annual appropriation for mosquito control nearly quadrupled. The Assembly and Senate authorized $290,000 of which the Mosquito Control Commission received $150,000, the air-spray program $100,000, and the experiment station "$40,000 for further expansion of research and education programs." To these funds, Governor Meynert added $110,000, of which $70,000 was earmarked for the construction of a new research laboratory at Rutgers. The new facility was named in honor of Thomas Headlee.[73]

Encephalitis outbreaks were not limited to California and New Jersey. Glen Collett and Don Rees reported in 1958 that there were 120 cases of encephalitis in Utah and 39 cases in Kansas. The next year medical entomologists explained a mysterious outbreak of encephalitis when they found the SLE virus in Florida. Epidemics recurred in Tampa and St. Petersburg in 1961 and 1962. As was the case in New Jersey, a team of CDC, Rockefeller Foundation, and Florida Board of Health entomologists discovered that a mosquito, *Cx. nigripalpus*, that had previously been considered an annoyance was the principal disease vector. Two years later Camden, New Jersey, and Houston, Texas, experienced St. Louis encephalitis outbreaks.

John Mulrennan used the St. Petersburg–Tampa Bay SLE episode to bolster support for mosquito control in Tallahassee. Since its high point in 1953, the Florida legislature had reduced appropriations for mosquito control. Mulrennan argued that the SLE and other arboviruses made it imperative that the Bureau of Entomology open two new research facilities. In 1962 the legislature funded an Encephalitis Research Center in Tampa and a mosquito control laboratory in Panama City.

Mosquito control underwent tremendous expansion in the fifteen years following the end of World War II. Fueled initially by concern over the possibility of the spread of mosquito-borne disease and the availability of miraculous insecticides, the anti-mosquito movement expanded across the nation. By the early 1960s there were signs of discontent within the crusader's ranks. Individuals like Provost and Peters questioned the reliance on chemical control. The 1959 New Jersey encephalitis epidemic demonstrated that chemical resistance had increased in the seven years since the California encephalitis epidemic. Too many mosquito programs continued to rely on chemicals. Provost and Peters believed they were destined for disappointment. "If the knife won't cut the butter," Provost declared, "it's not so much sharpening the knife as softening the butter that will render the task possible. The only way to soften our insect pests so they will respond better to our control tools is to learn their behavior. There is no other way."[74]

The 1960s introduced mosquito control to an even more formidable type of resistance. The publication of Rachel Carson's *Silent Spring* in 1962 marked the beginning of a new phase in the American environmental movement. Veteran mosquito warriors could only shake their heads in disbelief when they were attacked for doing too good a job of killing mosquitoes.

Chapter 9 Discontent and Resistance

Stan Freeborn liked to tell the story of how he found himself at the end of World War I standing outside a delousing station in Newport News. Freeborn watched as a line of soldiers passed through the facility. When one of the enlisted men paused, Freeborn asked him "how he felt now that he was all cleaned up." The soldier shook his head and declared, "Mister, I feel just plain lonesome."[1] Had Freeborn, who died in 1960, a year after retiring as UC Davis's chancellor, lived another decade, he might well have concluded that the mosquito crusaders had much in common with the deloused infantryman.

The 1960s represented a tectonic shift in the institutional standing of the mosquito and vector control movement. For seventy years the mosquito crusaders had prided themselves as the champions of public health and the first line of defense against the torment of winged blood-sucking pests. This changed in the tumultuous 1960s.

Many factors contributed to the public's disenchantment with mosquito control in the 1960s. The success of insecticides and water management in reducing nuisance and medically significant mosquitoes in the 1950s proved a double-edged sword. On the one hand, there was unprecedented growth in the scope of mosquito control. The expansion of mosquito control reached a high point between 1957 and 1963. Advocates of mosquito control during this period succeeded in launching new programs in the Midwest and along the Gulf Coast in Louisiana, Mississippi, and Texas. On the other hand, Clarence Cottam and his protégé, Rachel Carson, mounted a powerful attack on mosquito control.

The publication of Carson's *Silent Spring* in 1962 marked the beginning of the modern environmental movement in America. Eight years later on April 22, 1970, tens of thousands of Americans declared their commitment to environmental reform on the first Earth Day. For Dick Peters, Tommy Mulhern, John Mulrennan, and hundreds of other veteran mosquito warriors, Earth Day represented the end of the mosquito crusade and the beginning of a contentious debate about the value of mosquito control that continues to the present.

Few if any of the leaders of the American Mosquito Control Association (AMCA) anticipated the turmoil that would define the 1960s and early 1970s. There were numerous positive developments. New mosquito control programs formed. In 1958 organized mosquito control came to the Upper Midwest when the voters in six counties surrounding Minneapolis and St. Paul approved a ballot measure authorizing the formation of the Metropolitan Mosquito Control District.[2] Interest in mosquitoes and mosquito control in Minnesota had surfaced in the 1930s. In 1937 William Owen published a report on the state's mosquitoes as one of the University of Minnesota's technical bulletins. Twelve years later in 1950 Minnesota's assembly approved legislation enabling counties and municipalities to form mosquito control districts. Eight years elapsed before the formation of the first district.

In the early 1950s a young entomologist named Ralph Barr was one of the catalysts for the development of organized mosquito control in Minnesota. Barr, who earned his bachelor of science degree from Southern Methodist University and his doctorate in public health at Johns Hopkins University, joined the faculty of the University of Minnesota as an instructor in entomology and parasitology in 1950. During the next three years, Barr and his wife, Sylvia, traveled throughout the state collecting mosquitoes.

Barr left Minnesota in 1955 when he accepted a position at the University of Kansas. Three years later, the University of Minnesota published Barr's *Mosquitoes of Minnesota*. Sylvia Barr prepared the book's illustrations. "Mosquitoes," Barr observed in his introduction to the book, "are extremely important to the economy of the state of Minnesota. Their primary importance at the present time does not concern their ability to carry disease but rather lies in their nuisance value."[3] Surveys revealed that of the state's forty-nine species of mosquitoes, *Aedes vexans* comprised close to 80 percent of the mosquitoes captured in New Jersey light traps.[4] *Ae. vexans*, as its name suggests, is a particularly "irritating" pest. The public's desire to eliminate this "nuisance" led to the formation of the metropolitan district. When the mosquito commissioners appointed Albert W. Buzicky the district's first director, reporters joked that the commission had certainly chosen an aptly named individual to take the "buzz" out of St. Paul and Minneapolis.

A similar desire for relief from pest and nuisance mosquitoes motivated the formation of the Louisiana Mosquito Control Association (LMCA) late in

1957. Organized mosquito control in Louisiana grew out of the work of Edward S. Hathaway and Anderson B. Ritter. Hathaway, a retired professor of zoology at Tulane, and Ritter, who was the chief of the Vector Control Section of the Louisiana Department of Health, launched the campaign for mosquito control in the Bayou State in the 1950s. Hathaway was sixty-six years old when he joined the mosquito crusade. Born in 1886, Hathaway earned his doctorate in ichthyology and herpetology at the University of Wisconsin after World War I. Hathaway accepted a position at Tulane in 1925, and he remained there until his retirement in 1952.

Hathaway's retirement came at a propitious moment for those interested in advancing mosquito control in Louisiana. Public support for mosquito control surfaced in the late 1940s. The first call for mosquito control came from workers Freeport Sulphur Company (subsequently, Freeport McMoran Inc.) in Plaquemines Parish. "The employees," Louisiana State University entomologist Lamar Meeks noted in his history of the Louisiana control movement, organized a "grassroots effort to convince company officials to provide relief from pest mosquitoes." Frederick Deiler, a biologist and one of Hathaway's former students, suggested that the company ask Hathaway to prepare a mosquito survey. Hathaway identified the principal mosquito breeding areas in the salt marshes adjacent to the company. He recommended that the company and Plaquemines Parish pool their resources and mount a combined campaign. He warned that the problem extended beyond the parish's boundaries. Salt marsh and fresh water mosquitoes would remain a problem until Louisiana adopted a comprehensive control plan.[5]

Five years later, in 1953, angry citizens in New Orleans demanded relief from the broods of salt marsh mosquitoes that periodically descended on the Big Easy. When merchants complained that mosquitoes were driving tourists from the French Quarter's expensive restaurants and hotels, New Orleans' mayor deLesseps Morrison hired a local crop dusting company to spray DDT on the nearby "Little Woods" swamp. The spray treatments, however, failed to eliminate the mosquito problem. Public protests increased in the next eighteen months. In 1955 Mayor Morrison asked Hathaway and Ritter to join a blue-ribbon committee to study the city's mosquito problem. Morrison charged the five members of the New Orleans Metropolitan Mosquito Control Commission (MMCC) with devising a plan to protect the city against the mosquito blight.[6]

The commission was composed of Hathaway's former students and friends. In addition to Hathaway and Ritter, the MMCC included Rodney Jung, Fred Deiler, and Percy Viosca. Like Deiler, Jung, a New Orleans physician, was one of Hathaway's numerous former students. Viosca, the retired chief biologist for the Louisiana Wildlife and Fisheries Commission, was an old friend who shared Hathaway's passion for Louisiana's salt marshes and antipathy for mosquitoes.

Hathaway and Ritter were the guiding forces on the commission. The two men were a study in contrasts. Ritter, who was more than six feet tall, towered over Hathaway's diminutive frame. The men's temperaments were even more different. Ritter had an easy-going, relaxed attitude. In contrast, Hathaway had a reputation of approaching everything he did with zeal and discipline. There was nothing "laid back" about the little professor. At Tulane, Deiler remembered that he and his classmates "marvel[ed] at his [Hathaway's] dynamic demonstrations . . . and his boundless energy." An anonymous student declared in a course review that Hathaway was "crazy about system, precision, and organization, a (expletive deleted) slave driver but a square guy and a good teacher."[7]

Despite their differences, Hathaway and Ritter became close friends. They divided the commission's work between them. Ritter chaired the monthly meetings. Hathaway conducted the research. Early on the commission's members agreed that solving the salt marsh mosquito problem required a statewide effort. Hathaway's retirement allowed him to visit New Jersey, Texas, Arkansas, and Florida to observe how other states organized their anti-mosquito work. These trips convinced Hathaway that forming a statewide mosquito control association was the first step in organizing Louisiana's anti-mosquito campaign.

The LMCA came into being on December 2, 1957, with two objectives. The first was to win passage of a constitutional amendment that allowed parishes (counties) to form mosquito abatement districts. The second was to organize a research program with the goal of conducting a series of pilot studies comparing different methods of mosquito control.

Ritter drafted the Louisiana mosquito abatement law. Hathaway's conversations with mosquito control workers in Texas, Arkansas, and Florida guided Ritter's efforts. Both men believed that it would be an error to limit mosquito control to the coastal parishes. The Texas legislature had made this mistake when they formulated Texas's abatement law. Tommy Mulhern helped Jefferson County organize the state's first abatement district in 1949. Four years later Galveston County organized Texas's second district. At that rate, it would take sixty years before mosquito control reached Cameron County on the Mexican border.

By the mid-1950s the Texas mosquito crusaders were working to amend the state's abatement law. In 1955 representatives of Jefferson, Galveston, Orange, Chambers, and Brazoria counties met to discuss the revision of H.B. 127. There was general agreement that the current law was ineffective. At the end of the meeting, the delegates voted to form the Texas Gulf Coast Mosquito Control Association. Six years later this group changed its name to the Texas Mosquito Control Association.[8]

Hathaway outlined the LMCA's strategy for the development of mosquito control in Louisiana in a March 1958 speech in Atlantic City. "We bow in great respect," Hathaway told the delegates to the New Jersey Mosquito Extermination Association's forty-fifth meeting, "before the pioneering genius of John B. Smith and acknowledge our indebtedness to the New Jersey Mosquito Extermination Association for the service it has rendered during nearly a half a century of promoting research on mosquito biology and mosquito control." Hathaway explained that Louisiana's "infant mosquito control association" had formed in a "rather unorthodox manner." In other states, mosquito control work had preceded the formation of a mosquito control association. In Louisiana, Hathaway explained, "we formed the association first, and are now planning the necessary spadework."[9]

Hathaway and Ritter envisioned mosquito control in Louisiana developing in three stages. First, they planned to use the LMCA as a means of mobilizing support from cities, parishes, and businesses for mosquito research and pilot projects. Second, they hoped to win passage of a constitutional amendment in either 1958 or 1960 for the formation of mosquito abatement districts. Finally, after "progress in the pilot experiments seems to warrant the selection" of the best control methods "we must start campaigning for the setting up of mosquito control districts."[10]

The LMCA's first victory came in November 1958 when 52 percent of Louisiana's voters approved the constitutional amendment authorizing the formation of mosquito abatement districts. Funding research into mosquito biology and pilot programs proved more difficult. It was not until 1961 that Hathaway and Ritter were able to establish mosquito field stations along the Gulf Coast. The first pilot experiment in salt marsh mosquito control began in St. Mary's Parish in October 1961. When asked how long it would be before the state's first mosquito control district formed, Ritter counseled patience, explaining that it all depended on the pilot experiments. With good luck, he added, it might take five years. If the experiments did not pan out, it might be a decade.[11]

The critical moment in the development of mosquito control in Louisiana came in 1963 when a tremendous flight of *Ae. sollicitans* mosquitoes descended on Plaquemines Parish. Judge Leander Perez, whose position as head of the local political machine gave him near dictatorial power, ordered that something be done. One of Perez's subordinates hired a crop duster to spray the parish's salt marshes. The first treatment produced favorable results.

A few weeks later a second brood of mosquitoes rose from the marshes. This time the spray application failed. Fred Deiler, who had become the environmental specialist for the Freeport Sulphur Company, suggested that Judge Perez call Hathaway. In the conversation that ensued Hathaway told Perez he should hire John Beidler to make a mosquito survey and prepare a countywide mosquito control plan.[12]

By 1963 John Beidler had nearly twenty years of experience in waging war on mosquitoes. As a high school student Beidler had worked at the USDA Orlando laboratory. He took part in the first experiments using DDT against salt marsh mosquitoes in Brevard County near present-day Cocoa Beach. Later he went to Stuttgart, Arkansas, to work on an early test of DDT's potential as a means of residual control. After the war Beidler earned a degree in entomology at the University of Florida and went to work for John Mulrennan as one of the Florida Health Board's regional entomologists. Beidler became the director of the Indian River Mosquito Control District in 1955.

Retirement gave Hathaway time to visit Florida and see firsthand Brevard and Indian River counties' salt marsh impoundments. Hathaway was impressed. Charles Anderson, who later went on to become a leader in the LMCA, served as Hathaway's chauffeur on many of the professor's trips to Florida. Anderson retains a vivid memory of Hathaway carrying a stack of index cards listing the questions he wanted to ask. There was no escape once Hathaway started firing questions. Fortunately, John Beidler, Maurice Provost, and Jim Haegar always found time to answer Hathaway's inquiries.[13]

Hathaway did not hesitate when Perez asked for his recommendation in 1963. He told the judge that John Beidler was the man for the job. Perez ordered his secretary to contact Beidler and invite him to become Plaquemines Parish's mosquito consultant. Beidler agreed and spent several weeks surveying the parish. In November 1963 he presented the Parish's Police Jury (County Commission) with a plan for abating the salt marsh mosquito problem. Plaquemines Parish organized the state's first mosquito control district in April 1964. Parish officials tried to recruit Beidler to lead the district's operations. Beidler declined and recommended that the county interview Robert Barnett, the director of the Anastasia Mosquito Control District in St. Augustine, Florida. Barnett accepted the position.

During the next few years a stream of Florida-trained mosquito control workers made their way to Louisiana to provide their expertise in organizing the Bayou state's mosquito control program. George Carmichael, a Rutgers graduate who had worked for John Mulrennan in the Florida Health Board's Bureau of Entomology, guided the formation of Louisiana's second mosquito control district in Orleans Parish.[14] Carmichael, who had left Florida to organize Georgia's first mosquito control program in Chatham County, decided to remain in New Orleans and became the district's first director. Jefferson Parish hired Paul Hunt, director of the East Volusia Mosquito Control District, to serve as its consultant. By 1990 fifteen of Louisiana's sixty-four parishes had launched abatement districts.[15]

Hathaway and Ritter's work in Louisiana encouraged business leaders along the Mississippi Gulf Coast to form an anti-mosquito association. Since the 1925 *Ae. sollicitans* invasion that inspired Faulkner's novel and led to Congress's appropriation of fifty thousand dollars for the U.S. Public Health Service's (USPHS) salt marsh survey, Mississippians had repeatedly complained of periodic mosquito infestations. Flights of mosquitoes often reached towns nearly forty miles from the Gulf Coast. USDA researchers documented subsequent mosquito invasions in 1927, 1929, 1938, 1962, and 1963.[16]

The 1963 invasion led citizens in Hancock, Harrison, and Jackson counties to form a three county Gulf Coast Mosquito Control Commission. Fred Harden, who directed the St. Lucie Mosquito Control District in Florida in the late 1950s and served as the entomologist for the National Aeronautics and Space Administration's (NASA) Mississippi Test Operations (MTO) program, guided the development of the anti-mosquito work. He prepared a mosquito survey for the test operations area and the buffer zone that separated the federal facility from populated areas in Hancock, Harrison, and Jackson counties. Harden's survey confirmed that *Ae. sollicitans* mosquitoes were the principal problem. He reported that an independent Army Corps of Engineers study estimated that "work efficiency at the MTO due to mosquitoes was twenty-five percent less than that which could be normally expected, and that in addition, some crews left their jobs."[17]

Harden's anti-mosquito work in Mississippi was but the latest effort on NASA's part to prevent blood-sucking pests from interfering with the American space program. In September 1961, four months after President John F. Kennedy's pledge to land a man on the moon before the decade's end, NASA began a dramatic expansion of the Cape Canaveral launch facility. By the mid-1960s, NASA had purchased nearly ninety-thousand acres of land in Brevard and Volusia counties. Unfortunately, NASA's moonport was located in the center of one of the most pestiferous salt marshes along the Atlantic Coast.[18]

Evidence of NASA's unease about the potential mosquito problem surfaced late in 1961. In the fall Kurt Debus, who led launch operations at the Cape, Gen. Leighton Davis, commander of the Air Force Missile Test Center, and Florida's governor, Farris Bryant, formed a Joint Community Impact Coordination Committee (JCICC). The JCICC was charged with "recommending solutions, or a course of action, in connection with the problems that will have an impact" on both the space program and the communities surrounding the missile test project.[19]

Mosquitoes ranked high among NASA's concerns. Brevard County was notorious for its prodigious mosquito population. In the 1950s Brevard mosquito control workers documented "landing rate counts in excess of 500 mosquitoes on a person in a one minute interval." Fishermen complained of flights of mosquitoes that were so intense that they "extinguish[ed] the kerosene lanterns used by commercial fishermen working in the area after dark." In 1962 Maurice Provost sent a pair of researchers from the Entomological Research Center to collect mosquitoes at Cape Canaveral. Using hand nets the two men collected nearly four pounds of mosquitoes in a single hour.[20]

Jack Salmela, director of mosquito control in Brevard County, formulated the cooperative agreement between federal, state, and local authorities that safeguarded the space program from the mosquito menace. Salmela proposed that the Brevard Mosquito Control District would plan and construct a series of mosquito impoundments on the north Merritt Island launch area. NASA would fund the permanent control work that Salmela estimated would take three years to complete. NASA and Brevard County divided responsibility for temporary control measures.

By 1963 Salmela's teams had "eliminated breeding in 11,588 acres."[21] In the years after Neil Armstrong's brief walk on the moon, scientists and environmentalists praised Salmela for his contributions to the space program and environmental stewardship. "If it hadn't been for mosquito control," Jack Rogers, John Mulrennan's successor as Florida's state entomologist, maintained, "I don't know how the space program would have gotten off the ground."[22] In 1986 the U.S. Fish and Wildlife Service awarded "Jackie" Salmela its Conservation Service Award for his "care and perseverance" in protecting wildlife and the environment while controlling mosquitoes in Brevard County's salt marsh impoundments.[23]

Brevard and Indian River counties' successes in source reduction programs were not matched on Florida's Gulf Coast. Charlotte, Lee, and Collier counties in southwest Florida posed a particularly troublesome problem. Collectively, these counties contained hundreds of thousands of acres of salt marsh. Unlike the East Coast, State 2 Funds were not being utilized in southwestern Florida. As had been the case fifty years earlier in California, the chief reason for this was that local resources were insufficient to initiate control programs.

Sanibel Island was a prime example of the problems on Florida's southwest coast. Located in Lee County, Sanibel Island organized its own mosquito control district in 1953. When the Bureau of Entomology's Sanibel study ended, Maurice Provost left a plan for remedying the salt marsh mosquito problem. The difficulty was that there was an insufficient tax base on Sanibel to launch the control program. In 1955–1956, the Sanibel Mosquito Control District raised a total of $6,004.77 with a tax levy of 5.94 mills. The addition of the state aid funds meant that a total of $15,722.50 was available for mosquito control. This was a pittance. Source reduction required far more money than this.[24]

Maurice Provost had published an article in 1953 titled "The Watertable on Sanibel Island," which offered a model for how to permanently control the island's pest mosquitoes. Provost based his control strategy for Sanibel on the discovery that there were two main sources of mosquitoes on the island. The barrier island resembled an elongated saucer with a series of ridges surrounding its center. The swales between the ridges on the island's interior were the most troublesome source of mosquitoes. "The situation on Sanibel Island," Provost explained, "was such that surplus water was too slow in draining off the land, but once runoff was established, too much water escaped." Roadwork had exasperated the drainage problem. The construction of SR 867 (Periwinkle Way) along the island's length formed a "natural barrier" for runoff that created innumerable new breeding areas. The other sources of mosquitoes were the saltwater flats and mangrove swamps along Tarpon Bay. During the dry season the water that had accumulated in the interior swales dried up. When the summer rains came, the swales would become major breeding points.[25]

Provost recommended connecting the swales to form a slough running through the island's center. That would allow mosquito control to regulate the water level in the center of the island as well as permit predacious fish to move throughout the system and help control the production of mosquito larva. Provost believed that the lengthening and broadening of the central slough in the island's interior could be linked up into a single water-control system based at Tarpon Bay.

In 1957 T. Wayne Miller, Lee County Mosquito Control District (LCMCD) director, used State 2 Funds to purchase a dragline and went to work deepening the slough. Miller persuaded Fred Bishopp, who had retired in Fort Myers, to join the LCMCD's governing board. The two men proved to be a powerful team. Four years later the LCMCD had completed much of Provost's plan. The work represented "a giant step in the reduction of mosquito breeding on Sanibel" as well as "providing a continuous supply of fresh water to the residents on the island."[26]

Sanibel was one of the great success stories in the history of mosquito control. By the late 1960s Lee County Mosquito Control's impoundments on Sanibel had eliminated much of the salt marsh mosquito problem while vastly expanding the habitat for migratory birds and wildlife. Once considered one of the most pestiferous places on the planet, Sanibel became a mecca for shell collectors, birders, and sunbathers in the 1970s. In 1967 the U.S. Department of Interior designated 6,400 acres on Sanibel a wildlife refuge and bird sanctuary in honor of J. N. "Ding" Darling.[27] Three years later when Bishopp died LCMCD workers were nearing completion of a meandering wildlife drive built on top of mosquito impoundment dikes. Bishopp must have relished the irony that the J. N. "Ding" Darling National Wildlife Refuge depended on mosquito control. Today, nearly one million wildlife and bird enthusiasts annually visit the refuge.[28]

The mosquito crusader's accomplishments in Minnesota, Louisiana, Mississippi, and Florida, however, did not dampen the growing discontent with mosquito control. By the early 1960s, a new generation of environmentalists renewed Vogt and Cottam's attack on mosquito control. In the 1930s the anti-mosquito movement presented a unified response to the wildlife advocate's challenge. This was not the case in the sixties. By 1964 the mosquito crusaders disagreed over the methods and objectives of the control movement.

Ironically, the crisis within the control movement surfaced over the mosquito warriors' first success: the victory over the yellow fever mosquito. In 1900 the Reed Commission proved Finlay's hypothesis that the *Ae. aegypti* mosquito was the disease's vector. Gorgas's campaigns in Havana (1900–1902) and Panama (1904–1915) as well as Kohnke's and Porter's use of anti-mosquito measures in the 1905 New Orleans and Pensacola yellow fever epidemics demonstrated the importance of mosquito control. After World War II a number of veteran mosquito warriors called for a hemisphere-wide campaign to eradicate *Ae. aegypti* mosquitoes. Others within the mosquito crusade dismissed the *Ae. aegypti* eradication effort as unrealistic and pointless.

The debate about eradication came to a head in the late 1950s. Fred Soper was the leading advocate for mosquito eradication. Born in 1894 Soper earned his bachelor of science and master's degrees from the University of Kansas as well as a medical degree from Rush Hospital at the University of Chicago and a doctorate in public health from Johns Hopkins University. Soper worked from 1920 to 1942 for the Rockefeller Foundation in South America. During World War II Soper led the Allied effort against typhus epidemics in Algeria, Italy, and Egypt. At the end of the war Soper organized anti-malaria work in the Pacific theater. Between 1947 and his retirement in 1959 Soper led the Pan American Sanitary Bureau, which subsequently became the Pan American Health Organization.[29]

In 1920 the Rockefeller Foundation assigned Soper to the foundation's hookworm disease program in Brazil. During the next decade Soper led the

Foundation's anti-hookworm efforts in Brazil and Paraguay. Later Soper recalled that his work fighting hookworm had "taught him the value of detailed recording keeping, and the importance of continued visualization of accumulating data, and above all, the necessity of checking in the field the validity of operational reports." These skills proved invaluable in fighting the resurgence of yellow fever in South America in the 1930s.[30]

Many public health workers believed that yellow fever no longer posed a threat in the 1920s. This changed when yellow fever broke out in Rio de Janeiro in 1928. This episode was particularly disturbing because in the previous twelve months there had only been a single, clinically diagnosed instance of yellow fever in the entire Western Hemisphere. Outbreaks of the disease recurred in the next two years. Years later Soper admitted that he hesitated before accepting the foundation's offer for the assignment, which required "directing a service in which I had no direct experience."[31]

Soper spent his first three months learning everything he could about mosquitoes and mosquito control. Thirty years earlier William Gorgas had observed that in battling mosquitoes it was necessary to think like a mosquito. Soper modified this dictum declaring that in order to "get results with *aegypti*" you need to think like a "guarda" (mosquito control inspector). The first summer Soper lost twenty-seven pounds following "guarda" as they made their rounds up onto rooftops and through streets and alleys of innumerable towns and villages in northern Brazil. By early fall, he had formulated a plan for an *Ae. aegypti* eradication campaign.

Two years later in 1932 researchers discovered there were, in fact, two yellow fever transmission cycles. The urban cycle in which *Ae. aegypti* mosquitoes carry the virus to human beings was well understood. There was, however, a second transmission cycle involving another mosquito, *Haemagogus capricornii*, which transmitted the virus to monkeys.[32] The existence of this jungle cycle meant that as long as there was a population of *Ae. aegypti* mosquitoes in urban areas there would always be a threat of yellow fever epidemics.

In Brazil Soper faced formidable challenges in his war against *Ae. aegypti* mosquitoes in the 1930s. First, he had to overcome health official's skepticism of whether it was possible to eradicate an entire species of mosquitoes. Second, the Rockefeller Foundation's Yellow Fever Service covered only northern Brazil. This meant that even if Soper's limited campaign succeeded, there would be a constant threat of the reintroduction of *Ae. aegypti* mosquitoes to villages and towns where Soper's "guarda" had eliminated the vector.

The discovery that an exotic mosquito, *Anopheles gambiae*, was present in Brazil in the 1930s proved a catalyst for the eradication effort. This mosquito is the principal vector of malaria in Africa. Epidemiologists feared that if it became established it would lead to a tremendous increase in malaria throughout South America. These fears were realized in 1937–1938 when a malaria epidemic swept across Brazil. The malaria outbreak led the Brazilian government to organize the Malaria Service of Northeast Brazil in 1938. Soper and the Rockefeller Yellow Fever Service guided the Brazilian government in its efforts to eradicate the *An. gambiae*.

The critical breakthrough came early in 1939 when Marshall Barber visited Brazil. Barber had recently retired from the Rockefeller Foundation. Eighteen years earlier, Barber and Theodore Hayne had pioneered the use of Paris green as larvicide. Soper credited Barber with unraveling the puzzle of how to use Paris green to eradicate *An. gambiae*. Three days after his arrival Barber improvised a way of mixing the arsenic poison with dirt, gravel, or sand. Barber's discovery allowed workers to stockpile the Paris green mixture during the dry season. When the rains came, Soper's "guarda" were able to launch their campaign against *An. gambaie* when dust was no longer available. This resolved the principal logistic problem in using the larvicide. By 1940 Soper had succeeded in eradicating *An. gambiae* in six states and the Federal District in Brazil. By the mid-1940s Soper's "guarda" had driven *An. gambiae* out of South America.[33]

The success of the South American *An. gambiae* eradication measures generated support for expanding the initiative. During World War II, Soper served as a civilian member of the American Typhus Commission in the Mediterranean. When *An. gambiae* appeared in Upper Egypt, Soper recommended British health workers adopt an eradication campaign patterned on the Brazilian effort. Egyptian and British authorities initially dismissed Soper's recommendation. In 1944 a malaria epidemic raged along the Upper Nile. King Farouk asked the Rockefeller Foundation for assistance. Soper, who was fighting a typhus epidemic in Naples, Italy, agreed to take charge of the Egyptian *Gambiae* Eradication Service. Within a year, Soper's mosquito workers succeeded in driving *An. gambiae* out of Egypt.[34]

Soper returned to South America after the war and Soper took charge of the Pan American Sanitary Bureau (PASB) in 1947. At the PASB and its successor, the Pan American Health Organization (PAHO), Soper sought to expand the eradication effort. Bolivia had proposed a hemisphere-wide eradication effort as early as 1942. Five years later the PAHO's directors approved a resolution calling for all countries in the western hemisphere to join in a hemisphere-wide *Ae. aegypti* eradication effort. By 1961 twenty-five of the twenty-six member states had launched *Ae. aegypti* eradication programs. The United States was the single holdout.

Soper was unrelenting in his advocacy of the eradication concept. He presented the case for eradication at the 1958 annual meeting of the AMCA in Washington, D.C. Soper contended that "a series of technical, political and financial miracles" provided a basis for the success of the eradication concept. If the mosquito crusaders were resolute, Soper was confident that they could achieve total victory against disease-bearing insects.[35]

Soper based his faith in the eradication concept on his interpretation of the history of the anti-mosquito movement. Looking back over the course of the twentieth century, Soper contended that the anti-mosquito crusade had advanced in moments of national crisis. In World War I, the need to protect American servicemen had led the USPHS and military authorities to launch anti-mosquito work. During the Depression, the need to find work for millions of unemployed Americans led to a dramatic expansion of both anti-malaria and pest mosquito control work. Most recently, waging war in the Pacific and Europe had led to a dramatic expansion of the mosquito control movement.

Soper argued that World War II marked the beginning of a new epoch in the mosquito crusade. The discovery of new means of "chemical attack" provided the mosquito crusaders with the tools needed to win the final victory against mosquito-borne diseases. All that was lacking was the political will to employ these powerful chemical tools. Eradication of *Ae. aegypti* and anopheline mosquitoes was not a pipe dream. Soper called on the AMCA to throw their support behind the eradication effort.

A year later, Harold Gray delivered a thoughtful rebuttal to Soper's paean. Gray, who had retired as director of the Alameda Mosquito Abatement district in 1955, had nearly fifty years of experience waging war on mosquitoes. Like Soper, he based his argument against the eradication concept on his reading of history. Gray thought that Soper was overly optimistic. He faulted Soper for his confidence in the ability of insecticides to retain their effectiveness. Experience had taught Gray that resistance was a real and growing problem. Gray doubted that any chemical means of control would retain its usefulness for long.[36]

Two additional historical "contingencies" militated against the eradication concept. Even if Soper was correct and insecticide resistance did not become a problem, Gray argued that it would not be possible to sustain a global insecticide-based campaign indefinitely. Gray had observed firsthand the Depression's effect on mosquito control and public health initiatives. A worldwide economic downturn would inevitably lead to cutbacks in anti-mosquito work in poor countries. Moreover, he feared that if the Cold War should turn into a shooting war, the eradication effort would fall to the wayside. Eradication, Gray maintained, was a chimera, an unachievable goal. He maintained that a more realistic and attainable objective was seeking to control disease-bearing mosquitoes through a combination of naturalistic, mechanical, cultural, and chemical means.[37]

Figure 9.1. Harold Gray, circa 1949. Photo courtesy of Alameda County Mosquito Abatement District.

In the 1960s the fate of the eradication campaign was determined by historical events that had little to do with public health, entomology, and mosquito control. Fidel Castro seized power of Cuba in 1959. A year later Castro declared he was a communist and that he planned to make Cuba a socialist state. On April 17, 1961, a contingent of 1,500 Cuban exiles invaded Cuba at the Bay of Pigs. Three days later the U.S.-sponsored invasion ended in defeat. The majority of the CIA trained invaders were either dead or captured. The Bay of Pigs fiasco led to a shift in the Kennedy administration's Latin American policy. Kennedy pledged his support for an Alliance of Progress in which the United States would aid its Latin American neighbors.

Mosquitoes played a small but important role in shaping the new U.S. policy against Castro and winning Allies in the Americas. In 1961 the directors of PAHO reaffirmed their commitment to Soper's *Ae. aegypti* eradication campaign. One year later, Mexico's minister of health presented the U.S. surgeon general with a plastic case that contained "the last two *Ae. aegypti* mosquitoes in Mexico."[38] Critics of the United States pointed to the fact that in 1947 and again in 1961 the United States had committed itself to the eradication effort but no action had followed. President Kennedy announced in February 1963 that the United States would join the eradication effort. Kennedy declared that he was asking Congress for funds to launch an *Ae. aegypti* eradication program "to bring this country into conformity with the long established policy of the Pan American Health Organization to eliminate the threat of yellow fever in this hemisphere."[39]

The U.S. *Ae. aegypti* eradication program proved disastrous. Congress approved an initial appropriation of three million dollars for the eradication campaign in October 1963. Congress gave the Atlanta-based Centers for Disease Control (CDC) responsibility for directing the project. In the weeks before the bill's approval, officials at the CDC made it clear that they opposed the legislation. Public health historian Elizabeth Eldridge argues that government health workers fought against the eradication bill "tooth and nail." CDC officials maintained that there were more important public health needs. The last U.S. yellow fever outbreak had taken place in 1905. There had not been a single case of yellow fever in the United States since 1925. "Hardly anyone [at the CDC]," Eldridge contends, "saw the eradication of the *Aedes aegypti* mosquito . . . as a pressing health concern." The CDC's leadership believed that "the decision to eradicate *Aedes aegypti* mosquito was a political one, a matter of foreign policy."[40]

Donald Schliessmann, a sanitary engineer with experience in both the Tennessee Valley Authority (TVA) and Malaria Control in War Areas (MCWA), led the *Ae. aegypti* eradication program. Schliessmann spent his first six months organizing the eradication work. Between 1957 and 1961 the CDC's technology branch had conducted a pilot eradication effort in Pensacola. The Pensacola test demonstrated that eradication *Ae. aegypti* mosquitoes in nine southern states, Puerto Rico, the Virgin Islands, and Hawaii would take much longer than the five years authorized by Congress. More significantly, Schliessmann estimated that the total cost of the program would be close to 100 million dollars.[41]

The initial appropriation of three million dollars meant that Schliessmann was compelled to restrict the eradication effort to Florida, Puerto Rico, and the Virgin Islands in 1964. Schliessmann's plan was patterned on the MCWA effort. The attack on *Ae. aegypti* mosquitoes was divided into four phases: preparatory, attack, consolidation, and maintenance. During the preparatory phase, CDC inspectors would develop detailed *Ae. aegypti* surveys for the U.S. southeast, Puerto Rico, and the Virgin Islands. In the attack phase, the CDC planned to hire local workers to spray DDT. Insecticide treatments would taper off during the consolidation phase. In the maintenance phase CDC inspectors would conduct periodic surveys to make sure that *Ae. aegypti* mosquitoes had not gained reentry into treated areas.

Logistically, Schliessmann faced the problem of building an infrastructure to support the eradication effort. Securing power sprayers and trucks for the spray teams led to delays. When the eradication effort began in 1964, locally hired workers sprayed a 1¼ to 2 percent DDT emulsion on all of the premises in areas infested with *Ae. aegypti* mosquitoes. Spray teams repeated the treatments every ninety days.[42]

Russell Fontaine, who coordinated the University of California's mosquito research program between 1972 and 1986, was an early critic of the eradication effort. In 1965 Fontaine enumerated the program's weaknesses in a speech before the California Mosquito Control Association. Fontaine maintained that CDC planners had underestimated the magnitude of the *Ae. aegypti* problem. Hiring nine hundred contract employees to work on the spray teams had led to delays. Worse, the spray teams had proven inept. Fontaine's review of the CDC internal reports on the first year's work revealed that the "inspectors spraying fell short of acceptable standards during the first few months of the program." Fontaine concluded that the CDC must improve its supervision of the local effort.[43]

James Heidt, director of the Metro Dade County Mosquito Control Division, agreed with Fontaine. In 1966 Heidt used his presidential address to the Florida Anti-Mosquito Association to outline the problems in a speech titled "Problems that Confront Mosquito Control Organizations in Florida." By 1966 the *Ae. aegypti* eradication program was operating in seven counties in Florida. Heidt reported that "very little headway had been made" in large part because of "the vastness of the area, possible resistance to DDT, and insufficient equipment and personnel." The resistance problem was particularly worrisome. Heidt cautioned his fellow delegates not to forget "the lesson we should have learned from the misuse of the chlorinated hydrocarbons."[44]

Despite entomologists' and public health workers' misgivings, the Johnson administration continued to pump millions of dollars into the eradication effort. In 1965 Florida's Bureau of Entomology received $3,066,000 in federal support for its *Ae. aegypti* effort.[45] That year John Mulrennan reported to CDC officials in Atlanta that there was widespread resistance to DDT in Florida. USPHS funding for the *Ae. aegypti* eradication program leveled off at $3,364,656 in 1966 and $3,294,865 in 1967. A year later, weighed down with the mounting costs of the Vietnam War, President Johnson ordered "drastic reductions in expenditures" and asked Congress for a 10 percent surtax on income taxes.[46] After four years of work and more than twelve million dollars of expenditures, the federal government terminated the *Ae. aegypti* eradication effort.[47]

Ae. aegypti mosquitoes were not the only creatures showing resistance to DDT and pesticides. In June 1962 the *New Yorker* published the first in a series of articles by Rachel Carson describing the environmental damage of pesticides in general and DDT in particular. Three months later the articles were published in book form. Most historians consider Carson's *Silent Spring* to be the beginning of the modern environmental movement. *Silent Spring* played an important role in the creation of the Environmental Protection Agency (EPA) eight years later in 1970, the banning of DDT in 1972, and Congress's approval of the Endangered Species Act in 1973.[48]

Silent Spring appeared at a moment when the American public was keenly sensitive to environmental and health issues. At least this is the argument that Carol Anelli, Christian Krupke, and Renée Prasad presented in their 2006 analysis of the public and scientific communities' response to *Silent Spring*. Anelli, Krupke, and Prasad contend that a series of events in the 1950s and early 1960s sharpened Americans' concern about the environment. Public concern about the environment grew in 1959 after a Food and Drug Administration (FDA) recall of herbicide-laced cranberries. Two years later newspapers published the reports of a national "Baby Tooth Survey" that revealed the presence of Strontium-90, fallout from atmospheric nuclear tests, in infants' teeth. A year later in July, one month after the first installment of Carson's article appeared in the *New Yorker*, the FDA successfully prohibited an American pharmaceutical company's efforts to introduce the drug thalidomide. Pictures of infants in Europe and Great Britain suffering horrific birth defects confirmed the public's worst fears. A growing number of Americans began to doubt government scientists and government officials who promised "miracle drugs," insect-free crops, and mosquito-less nights.[49]

Silent Spring appeared in this context. Carson's great talent lay in her ability to create a series of evocative images of how a cabal of pesticide salesmen, applied entomologists, and government workers had fired a "chemical barrage" against the "fabric of life." In the closing sentences of *Silent Spring*, Carson directed her full fury at the arrogant "practitioners of chemical control" who seek "the control of nature" through their "Stone Age science." "It is our alarming misfortune," Carson declared, "that so primitive a science has armed itself with the most modern and terrible weapons, and that in turning them against the insects it has turned them against the earth."[50]

Carson's *Silent Spring* provoked an immediate reaction. Respected associations and organizations such as the American Medical Association, the American Nutrition Foundation, and the USDA Agricultural Research Service criticized Carson's findings. Three months after the book's publication, Don Collins, editor of the American Mosquito Control Association's journal *Mosquito News*, praised and criticized Carson's work. "It is unfortunate that so many critics have been uncritical," Collins observed, "and, by confusing the meritorious with the meretricious, have come close to obscuring the merits of the book—for it does have merits. The author has done a real service in bringing together in one place a great deal of information of the dangers of improper use of pesticides, the necessity for regulating their distribution and application, on the desirability of using biological control measures whenever possible, and, above all, in emphasizing the need for research."[51]

Critics outside the mosquito control community were even harsher. Anelli, Krupke, and Prasad report in their review of the scientific response to *Silent Spring* that Carson was particularly dismayed with William J. Darby's review. Darby, who was the chair of the biochemistry department at Vanderbilt, "found no merit in the book" predicting that Carson would find ardent supporters among "organic farmers, the antifluoride leaguers, the worshippers of 'natural foods,' those who cling to the philosophy of vital principle, and pseudo-scientists and faddists."[52]

Carson's opponents misunderstood the book's significance. Carson had written *Silent Spring* for a far greater audience than academic, professional, and government readers. Her objective was to awaken the public to the dangers of what she considered the wholesale misuse of pesticides. Her critics and supporters were in agreement that she succeeded in this. Six months after the book's publication, President Kennedy convened a White House conference to examine the use of pesticides.

Mosquito control fell under heavy criticism in the debate about *Silent Spring*, which was ironic for several reasons. First, the growing problem of chemical resistance had forced most mosquito control districts to stop using DDT. Moreover, since the late 1940s individuals like Harold Gray, Edward Knipling, Dick Peters, and Maurice Provost had cautioned against what they considered the excessive reliance on insecticides. By 1959 four University of California entomologists (Vern Stern, Ray Smith, Robert van den Bosch, and Kenneth Hagen) published a seminal paper advocating the "Integrated Control Concept." This paper formed the basis of what later became known as integrated pest management (IPM).

The *Ae. aegypti* eradication program added to the public's disillusionment with mosquito control. As politicians and public commentators debated *Silent Spring*, the CDC continued to dispatch hundreds of workers to spray DDT in urban areas in Florida, Texas, Louisiana, Mississippi, Alabama, and Georgia. The irony was that the majority of organized mosquito control programs in these states had abandoned the use of DDT because of the resistance problem.

In April 1966 a citizen group on Long Island filed a brief before the New York State Supreme Court asking for a permanent ban of the use of DDT by the Suffolk County Mosquito Control Commission (SCMCC). The previous year local opposition to DDT and a growing resistance problem convinced Christian Williamson, the SCMCC's superintendent, to phase out DDT from the control program. Williamson reported that "in the course of the summer of 1965 we had eliminated DDT from all of our program with the exception of the treatment of corner catch basins and drainage sumps."[53]

A year later, at the beginning of August 1966, the Suffolk mosquito commission announced that the district was no longer using DDT. Williamson's voluntary measures did not satisfy the commission's opponents. In mid-August, Suffolk county environmental activists won a temporary injunction from the New York Supreme Court prohibiting the mosquito commission's use of DDT pending the court's ruling on the insecticide.

Dennis Puleston, one of the leaders of the Suffolk county environmentalists, credited "Rachel Carson's powerful and eloquent indictment of the chlorinated hydrocarbon pesticides" as sparking the group's action. For several years Puleston had noted a steady decline in the osprey and blue claw crab population. "It was not difficult," he later declared, "to point a finger at the chief culprit. For the past 20 years, the Suffolk County Mosquito Control Commission had been aerially spraying DDT in its largely unsuccessful efforts to reduce mosquito populations."[54]

The New York Supreme Court justices heard the arguments in the environmental class action against Williamson and the Suffolk County Mosquito Commission in November 1966. The court issued its ruling a year later in December 1967. During this period the court's temporary injunction barring the Suffolk mosquito commission from using DDT remained in force. In its December judgment, the Court ruled in favor of the Suffolk mosquito commission noting that the "benefit obtained by the control and elimination of mosquitoes, and the relatively minimal upset of the balance of nature without undue harm or danger to man or beast through the use of DDT, is a valid governmental operation."[55]

The court's ruling proved damaging. To the public, the Suffolk mosquito commission appeared to be advocating the continued use of DDT. The commission's position, however, was more nuanced than this. Williamson and the Suffolk mosquito commissioners viewed the injunction as an attempt to circumvent the New York legislature's "properly delegated authority" that mandated the commission's control of mosquitoes. To the commission, the use of

DDT was not the issue. The Supreme Court agreed and affirmed the "propriety of the commission's utilization of a particular method of implementation of a properly delegated authority." The judges ruled that the plaintiff's claims were too "nebulous to warrant intervention." The court suggested that the remedy to this situation might be "more properly located with the policy making legislative unit as opposed to the policy executing body."[56]

There was considerable irony in the Suffolk environmentalists attack on Christian Williamson. Williamson had served as the Suffolk mosquito commission's superintendent since 1936. Under Williamson and his brother John's leadership, the Suffolk commission had developed a solid program after its uncertain start in 1933. The Williamsons worked to correct the problems created by Civil Works Administration (CWA) work relief program. In March 1938 Clarence Cottam had singled Williamson and J. Lyell Clarke for their "successful coordination of wildlife conservation and mosquito control." Cottam pointed to Williamson's construction of "valuable wildlife ponds" as an example of the "outstanding cooperation" between advocates of mosquito control and conservation groups.[57]

The Suffolk court case was the beginning of a series of legal efforts to curtail the power of mosquito control districts. Dennis Puleston and his allies were "emboldened" by the fight with Suffolk mosquito commission. In the fall of 1967 the Suffolk group decided to organize themselves as the Environmental Defense Fund (EDF). Five years later in 1972 Puleston, then chairman of the EDF's board of directors, celebrated the federal ban on DDT.

While Puleston and the EDF were organizing a nationwide campaign against DDT, "the environmental 'bug' . . . hit the State of Illinois." In August 1968 a citizen committee appointed by the Village of Palatine (near Chicago) called on the Northwest Mosquito Abatement District (NMAD) to "eliminate malathion from the District's adulticiding program." Six months later, the trustees of the Village of Palatine passed an ordinance making "it unlawful for any person, firm or corporation to operate mist sprayers or other types of spraying and fogging devices which introduce into the atmosphere chlorinated hydrocarbons or organophosphates within the corporate limits of the Village of Palatine."[58]

After consulting with the district's attorney, the NMAD's trustees adopted a "wait-and-see status while all legal aspects were being investigated." Two years later the Village of Schaumberg passed a similar ordinance prohibiting the NMAD from undertaking anti-mosquito work within the village's city limits. In August 1971 the NMAD filed suit against both Palatine and Schaumberg. In 1973 the Cook County Circuit Court ruled in favor of the mosquito control district. The Circuit Court ruling affirmed the right of the State of Illinois to "restrict the use of pesticides more narrowly than the federal government." Local governments, however, "were precluded from all jurisdiction and authority over pesticides and the regulation of pesticides."[59]

By the early 1970s the mosquito crusaders were in retreat. The mosquito warriors' accomplishments meant little on Earth Day. Demonstrations took place throughout the country. In Fort Lauderdale, the Florida Anti-Mosquito Association held its annual meeting. John Mulrennan expressed the frustration felt by many of the veteran mosquito fighters. Mulrennan declared that he was tired of hearing "'Johnny-come-latelys' who have all the answers to the past and future environmental problems of this state" prattle about the environment. Mulrennan warned that individuals who knew nothing about mosquito control "will try to run rough shod if given an opportunity" and do irreparable harm to mosquito control.[60] Mulrennan's comrades in California, Louisiana, New Jersey, and Utah agreed. After Earth Day, the mosquito warriors felt "just plain lonesome."

Epilogue

The End of the Crusade

The day is gone when we can improve our relations with fellow man or fellow animals by simply reaching for a gun. We need to understand, and for this we must learn. And so we need only learn, learn, and learn some more to appreciate that our natural environment is full of wonder and beauty, even if we learn about some practical phenomenon as the common milieu of birds, mosquitoes, and viruses.

—MAURICE PROVOST

PERHAPS YOU SHOULD CONSIDER changing your name," Grant Walton, director of the Division of Environmental Quality in the New Jersey Department of Environmental Protection (DEP) told the participants in the 1973 meeting of the New Jersey Mosquito Extermination Association (NJMEA). "You've been in business for sixty years and haven't exterminated all the mosquitoes yet, people may question what's been going on all those years. . . . I can foresee in a year or so you will have the New Jersey Mosquito Preservation Society, which will be holding meetings to counteract yours."[1] Walton's banter was not entirely in jest. A year later Henry Rupp, a former English professor who left the lecture hall to join the mosquito crusade, announced "that this issue of the Proceedings of the New Jersey Mosquito Extermination Association is the final record of this Association, which in July of 1974 voted to reorganize itself as the New Jersey Mosquito Control Association, Incorporated."[2]

Names are important. They reveal much about how individuals, institutions, and associations view themselves. On April 22, 1970, Earth Day, the

New Jersey legislature organized the Department of Environmental Protection out of the old Department of Conservation and Economic Development (DCED). Grant Walton, who went on to become the director of the agricultural experiment station, was the first in a series of DEP administrators prepared to question the goals and methods of the state's nineteen mosquito commissions. The New Jersey mosquito warriors' decision to drop the word "extermination" is a useful point to mark the end of the mosquito crusade. Spencer Miller, Thomas Headlee, and Ralph Hunt had two goals when they organized the NJMEA in 1913. Their immediate objective was to defend the law authorizing the formation of mosquito commissions. Miller, Headlee, and Hunt feared that politicians would undermine the nascent mosquito control movement. Their long-term objective was eliminating the mosquito pest.

Sixty-one years later, Miller, Headlee, and Hunt's successors conceded that extermination was an unrealizable and undesirable goal. The name change was a declaration of an armistice in the association's take-no-prisoners war of extermination against the mosquito pest. In a sense, the decision represented a return to John Smith's original vision of the mosquito crusade. "I can see no reason why," Smith observed in 1901, "New Jersey mosquitoes should not be reduced to such a point as to be practically unnoticed. Extermination, be it noted, is *not* claimed. The plan contemplates permanent relief, based upon intelligent activity, along well developed lines."[3]

California and Florida followed New Jersey's lead in changing the names of their associations. In 1976 the California Mosquito Control Association expanded its responsibilities to include ticks, fleas, and other disease-bearing insects to become the California Mosquito and Vector Control Association (CMVCA).[4] The Florida Anti-Mosquito Association (FAMA) retained its name for another decade. John Beidler, a past president of the association, explained the eventual name change at the 2006 Louisiana Mosquito Control Association's annual meeting. "The association's members got tired of checking into hotels who had put up signs welcoming the conference attendees saying 'Welcome Florida Anti-Mosquito Control Association.' We were not," Beidler tersely added, "into that."[5]

The 1970s marked the passing of a generation of mosquito warriors who had led the movement since the Depression and World War II. Bailey Pepper, Fred Bishopp, and Willard King died in 1970. Don Collins, the highly respected editor of *Mosquito News*, died in 1973. In 1976, Don Rees suffered a fatal heart attack on his way home from a doctor's visit at which he had received a clean bill of health.

Other veterans like John Mulrennan and Dick Peters retired. Mulrennan, who guided the development of Florida's anti-mosquito movement, believed that the "golden era of mosquito control" had passed. "The future can be bright, but it is going to take a greater job of informing the lawmakers, and other public officials, and the general public." Mulrennan, on the eve of retirement, admonished his fellow mosquito warriors to maintain high standards. "There is an old and simple adage," Mulrennan declared,

> "your sins will find you out." There have been occasions where local operations have not carried out recognized and approved mosquito control practices. Such instances have a detrimental effect on all mosquito control operations. There is a right and a wrong way to carry out an operation, but it seems, today there are those who feel they can carry out actions not recognized as legitimate, and in some instances may be over the borderline. Under these circumstances, the image of mosquito control is suspect and will be difficult to defend in the future.[6]

Mulrennan and Peters believed that there was a critical need for research in the coming "third era of mosquito control." Both men were dismayed when politicians in Tallahassee and Sacramento in the 1960s slashed funds for research and dismantled well-established research programs. In May 1965 a California legislative committee ordered the Bureau of Vector Control (BVC) to close its operational research program in Fresno. In Florida Mulrennan faced the daunting challenge of maintaining the state board of health's Vero Beach and Panama City mosquito research facilities. In 1970 the Florida legislature dissolved the state board of health and created Department of Health and Rehabilitative Services (DHRS). Mulrennan feared that under the DHRS, state support for mosquito control would dwindle.

In California Richard Frolli, manager of the Kings Mosquito Abatement District, echoed Mulrennan's concerns. Like Mulrennan, Frolli believed that the mosquito crusade had come to a turning point.

> These are crucial times. Our pesticides are failing! Our basic solutions for mosquito control are dying! . . . We must change our basic strategy, we must change our basic solution, we must change our district images to one other than spray districts if we are to be effective in mosquito abatement. . . . We have a crisis in mosquito control. The crisis is a mental one—a reluctance to accept change. But this crisis will break as a fever does. Someday, we'll all realize and the public as well, just how ridiculous the dependence upon repetitive spraying is as we get a clear grasp of the new realities of mosquito abatement, and define clearly new values and priorities.[7]

Unfortunately, while Mulrennan was fighting to save Florida's mosquito research program and Frolli was admonishing the mosquito crusaders to accept the "new realities" of mosquito control, a scandal threatened the future of one of the nation's first mosquito programs. In December 1970 the *Newark Star-Ledger* reported on its front page under the headline "Mosquito Agency Corrupt" that the New Jersey State Commission of Investigation (SCI) had accused Harry Callari, executive secretary of the Hudson County Mosquito Commission, of bribery and fraud.[8] The SCI's hearing revealed that corruption was rank in the Hudson Mosquito Commission. Callari, who served on the state mosquito commission, and David Strauss, the Hudson commission's treasurer and past president of the New Jersey Mosquito Extermination Association, orchestrated a series of kickbacks and payoffs that had netted them nearly three quarters of a million dollars.[9]

The transcript of the SIC's hearing on the corruption charges against the Hudson Mosquito Commission is riveting. David Strauss acknowledged that he regularly signed checks and vouchers without reading them. Strauss's defense for this practice was that he was legally blind and signed whatever was given to him.[10] After being given immunity from prosecution, Joseph Rossillo provided a detailed description of how Callari and Strauss received kickbacks from fraudulent vouchers. In one instance, Rossillo received a check for $25,000 for excavating nonexistent drainage ditches. He told the commission that he had deposited the check in his bank account. Callari and Rossillo had agreed to a 75/25 split. Three days after being paid for the bogus work, Rossillo returned to the Hudson Mosquito Commission office with $18,800 dollars in "thousand dollar packages, hundred dollar bills . . . wrapped up in rubber bands, wrapped up into an envelope and then wrapped up with rubber bands." Rossillo carried the money in his attaché case to Callari's office. The SCI investigator asked

> Q: What did you do with the package?
>
> A: I then opened up my attaché case; took the package out; unwrapped the rubber bands; took the money out of the envelope; took the rubber bands off; laid it on the table; counted it and gave it to Mr. Callari.
>
> Q: And what did he say?
>
> A: "Thank you." We counted it out, okay. He asked me how I was, how my father was.[11]

One year later, in December 1971, Callari pleaded guilty to embezzling $150,000. The court sentenced Callari to serve two to four years in the state prison.

Hudson County's troubles multiplied after Callari's confession. Subsequent investigations revealed that Callari and others had used Hudson County parks as a landfill. In September 1972, a grand jury indicted eleven individuals including the Hudson County's former Democratic chairman and county freeholder (commissioner), John J. Kenny, for using county land as "a private garbage dump."[12]

The Callari scandal proved fatal for the Hudson County Mosquito Commission. In March 1971 the Hudson freeholders dissolved the independent commission. A year later the New Jersey legislature gave counties the option of reorganizing their governments. Under the Optional County Charter Law, voters in each county could appoint a study commission. Hudson, Atlantic, and Mercer counties adopted an elected county executive form of government. Union County voters approved a county manager system. Most local boards, like the mosquito commissions, disappeared under the new county governments.[13]

In the midst of these setbacks, the third era in the mosquito control movement was beginning. During the 1970s and 1980s a series of scientific and technical developments revolutionized mosquito abatement and medical entomologists' understanding of the transmission of diseases caused by mosquito-borne pathogens. Researchers made important advances in both surveillance of mosquito populations and the isolation of pathogens in mosquitoes. Simultaneously, scientists in Israel and the United States developed bacterial organisms that offered an ecologically sound means of biological control. In 1983 the Environmental Protection Agency (EPA) approved *Bacillus thuringiensis israelensis* (*Bti*) for use as a mosquito larvicide. Eight years later the EPA registered *Bacillus sphaericus* (*B. sphaericus*) for control of mosquito larvae. These bacteria produce toxin that disrupt the gut of mosquito larvae with no effect on human beings. Subsequent EPA testing has demonstrated that "microbial larvicides do not pose risks to wildlife, non-target species, or the environment when used according to label directions."[14]

Equally important research on marsh ecology led to changes in mosquito control. In 1982 the U.S. Department of Commerce Coastal Zone Management (CZM) program funded an eight-year study of the effects of different water management practices on biologically sensitive wetlands. The CZM study concluded that rotational impoundment management (RIM) offered a means of protecting natural resources while maintaining control of mosquitoes. In Florida, this study led to the widespread adoption of the RIM technique throughout the state's saltwater impoundments.[15]

Despite these developments, mosquito control remains controversial. Public opposition to pesticides has intensified. Environmentalists continue to voice their concern about mosquito control's impact on biologically sensitive areas. At same time, the control movement faces new challenges from mosquitoes and pathogens. At the beginning of the twentieth century, entomologists, parasitologists, and epidemiologists such as Howard, Smith, and Herms thought it possible to differentiate between nuisance and disease-bearing mosquitoes. Today, entomologists have discovered that mosquitoes once considered only nuisances are, in fact, capable of transmitting disease pathogens. In the eastern United States, *Culex salinarius* was considered only a nuisance until the arrival of West Nile virus in 1999. Entomologists and epidemiologists are at work unraveling the virus vector–human–relationship. The future of the mosquito control movement in the twenty-first century depends on understanding this relationship.[16] John Smith was right. Science must guide mosquito abatement. In a time of decreasing research budgets, the challenges facing mosquito control workers are formidable.

The changing character of the American anti-mosquito crusade is most clearly shown in the movement's leaders. A new generation of entomologists, engineers, and mosquito control workers emerged to lead the anti-mosquito movement. Judy Hansen is perhaps the best representative of this group. In 1959 Hansen, a single parent with three children, was hired as a part-time worker at the Cape May County Mosquito Commission. Hansen (whose talent as a tap dancer led to an appearance on Ted Mack's *Amateur Hour*) quickly mastered the essentials of mosquito identification. In 1963 Hansen became a full-time employee at the mosquito commission. During the next decade, she regularly drove to Rutgers where Daniel Manley Jobbins and Lyell Hagman guided her study of taxonomy and water management.

In 1973 Boyd Lafferty, superintendent of the Cape May Mosquito Commission, died during open-heart surgery. The commissioners asked Hansen to lead the mosquito program. They offered her the position of the mosquito commission's executive director. Hansen refused. She later explained that she had three children to support. Under the Cape May charter, an executive director was a political appointment, not a civil service position. As a single parent, she needed a secure post. The commissioners reconsidered their decision and agreed to make her New Jersey's first woman mosquito superintendent.[17]

Judy Hansen transformed the Cape May Mosquito Commission. The district office was housed in an old World War II prisoner-of-war camp. When she took charge, termites were chewing through the walls. Under Hansen's leadership, Cape May became known as a center for progressive mosquito control practices. Twice chosen to lead the New Jersey Mosquito Control Association, Hansen served in 1989 as the American Mosquito Control Association's first woman president.

Judy Hansen was a testament that the vision of John Smith and William Herms had continued. The "golden era" of the mosquito crusade had ended. Nevertheless, the serious work of protecting the public's health and comfort remained. In her presidential address titled "Our Race to the Future," Hansen reflected on the history of the anti-mosquito movement. "The 1960s," Hansen observed,

> was the beginning of change, a time of unrest in this country. *Silent Spring* by Rachel Carson was published. Earth Day 1970 marked the beginning of environmental awareness in the form of laws and regulations. Some of these laws and regulations were long overdue and some were unreasonable. Suddenly it was fashionable to be an environmentalist. I resent this in many ways because I feel we in mosquito control were the first environmentalists when it was not the thing to be. We cleaned up after careless builders moved out of a newly developed area and eased the drainage problems that caused mosquito breeding. We established ponds and ditches in marshes and uplands that were beneficial to wildlife and increased productivity on marshes. We were the first biologists who knew the value of a marsh and that it was not just a vast wasteland to be filled. . . . This is a success story. Many of our distinguished colleagues of the past managed to get their message across. It is now our responsibility to carry the message to the future.[18]

For Hansen, Herms's farewell imperative rang true. Future mosquito warriors must seek to "*know well* the insects" and "*keep faith* with the folks who look to you for the accomplishments of mosquito control" [emphasis in the original]. Individuals like Smith, Herms, Freeborn, Rees, Mulhern, Peters, Reeves, Provost, and Hathaway acted well their parts. Their message was simple and direct. "You have a job to do—do it with sincerity and devotion."[19] It would be foolish to forget this.

Notes

Abbreviations

JBSs	Papers and Letters of John B. Smith. Office Books of Letters Sent 1900–1912, Department of Entomology, Blake Hall, Rutgers, New Brunswick, New Jersey.
JBSr	Papers and Letters of John B. Smith. Office Books of Letters Received 1900–1912, Department of Entomology, Blake Hall, Rutgers, New Brunswick, New Jersey.
MN	*Mosquito News*
NYT	*New York Times*
PCMCA	*Proceedings of the California Mosquito Control Association*
	The California Mosquito Control Association changed its name to the California Mosquito and Vector Control Association (CMVCA) in 1977; the CMVCA became the Mosquito and Vector Control Association of California (MVCAC) in 1996.
PCMVCA	*Proceedings of the California Mosquito and Vector Control Association*
PMVCAC	*Proceedings of the Mosquito and Vector Control Association of California*
PFAMA	*Proceedings of the Florida Anti-Mosquito Association*
	The Florida-Anti Mosquito Association changed its name to the Florida Mosquito Control Association (FMCA) in 1990.
PNJMEA	*Proceedings of the New Jersey Mosquito Extermination Association*
	The New Jersey Mosquito Extermination Association changed its name to the New Jersey Mosquito Control Association in 1974.
PUMAA	*Proceedings of the Utah Mosquito Abatement Association*

Introduction

1. "First Direct Dial System across Nation," *Alameda Times Star*, November 11, 1951, 1.
2. Mrs. Harvey Bein, "Holding the Line," *PNJMEA* 39 (1952), 39.
3. "Malaria in Washington," *NYT*, July 16, 1879.
4. Quoted in John Mulrennan Sr. "Traveling Through Memory Lane of Mosquito Control in Florida," *PFAMA* 47 (1976), 3.
5. Samuel Rickard Christophers, *Aedes Aegypti(L.): The Yellow Fever Mosquito* (Cambridge: Cambridge University Press, 1960), 2.

6. Donald Johnson traced the origin of the word "mosquito" to India in the *Atharva Veda*. In classical Sanskrit the word for "mosquito" became "masaka." The Roman use of "musca" (fly) comes from Sanskrit. See Donald R. Johnson, "Why "Mosquito?"" *Mosquito News* 25, no. 3 (1965).
7. Quoted in Austin W. Morrill Jr., "News and Notes," *Mosquito News* 22, no. 2 (1962): 216.
8. Christophers, *Aedes Aegypti (L.)*, 2.
9. The Louisiana Mosquito Control Association chose "Les Maringouin" as the title for the Association's newsletter.
10. Federal Writers' Project, *New Jersey: A Guide to Its Present and Past*, American Guides Series (New York: Viking Press, 1939), 80.
11. John Hannon, "Great Mosquito Wars," *Rutgers Magazine*, July–September, 1989.
12. Ralph Hunt, "The Anti-Mosquito Movement," in *PNJMEA* 2 (1915): 10.
13. Leland Howard, Harrison G. Dyar, and Frederick Knab, *The Mosquitoes of North and Central America and the West Indies*, vol. 1 (Washington, D.C.: Carnegie Institution of Washington, 1912), 9.
14. Erwin H. Ackerknecht, *Malaria in the Upper Mississippi Valley 1760–1900* (Baltimore: Johns Hopkins University Press, 1945), 97.
15. Quoted in Howard, Dyar, and Knab, *Mosquitoes of North and Central America and the West Indies*, 11, 12.
16. Samuel Eliot Morison, *The Oxford History of the American People* (New York: Oxford University Press, 1965), 49.
17. Margaret Humphreys, *Malaria: Poverty, Race, and Public Health in the United States* (Baltimore: Johns Hopkins University Press, 2001), 24.
18. Quoted in Moses Coit Tyler, *A History of American Literature* (New York: G. P. Putnam's Sons, 1879), 123.
19. Colin G. Calloway, *New Worlds for All: Indians, Europeans, and the Remaking of Early America* (Baltimore: Johns Hopkins University Press, 1997), 76.
20. Herbert Samworth, "John Eliot and America's First Bible," Sola Scriptura, http://www.solagroup.org/articles/historyofthebible/hotb_0005.html.
21. Jonathan Dickenson, *Jonathan Dickinson's Journal or God's Protecting Providence* (Stuart, Florida: Southeastern Printing, 1981), 20. See Gordon Patterson, *The Mosquito Wars: A History of Mosquito Control in Florida* (Gainesville: University Press of Florida, 2004), 9–11.
22. William Byrd and Edmund Ruffin, *The Westover Manuscripts: Containing the History of the Dividing Line Betwixt Virginia and North Carolina; a Journey to the Land of Eden, a.d. 1736: And a Progress to the Mines* (Petersburg, Va.: Printed by E. and J. C. Ruffin, 1841), 77.
23. Gregg Cantrell, *Stephen F. Austin, Empresario of Texas* (New Haven: Yale University Press, 1999), 137.
24. Jerald T. Milanich, *The Timucua*, The Peoples of America series (Oxford, UK; Cambridge, Mass.: Blackwell Publishers, 1996), 32.
25. Leland Howard, *Mosquitoes: How They Live; How They Carry Disease; How They Are Classified; How They May Be Destroyed.* (New York: McClure Phillips & Co., 1901), 41–42.
26. Austin. W. Morrill Jr., "News and Notes," *MN* 19, no. 1 (1959): 125. See Joshua Wolf Shenk, "The True Lincoln," *Time* 166, no. 1 (2005): 44.
27. Jason Zasky, "Get Your Wings: On the Centennial of First Flight, Rediscover the Remarkable Achievements of the Wright Brothers," http://www.failuremag.com/arch_science_wright_brothers.html.
28. Ibid.

29. Venita Jay, "Sir Patrick Manson: Father of Tropical Medicine," *Archives of Pathology and Laboratory Medicine* 124, no. 11 (2000): 1595.
30. B. F. Eldridge, "Patrick Manson and the Discovery Age of Vector Biology," *Journal of the American Mosquito Control Association* 8, no. 3 (1992): 216, 217. Manson's belief that mosquitoes took a single blood meal and "produced a single batch of eggs" convinced him that "filarial larvae escaped into the water from the bodies of dead mosquitoes and that human hosts were infected by drinking water infested with filarial larvae." Manson's error, Bruce Eldridge rightly observed, "in no way minimizes the importance of his monumental discovery."
31. Taxonomists have employed a number of different names for "yellow fever" mosquito before settling on *Aedes aegypti*. In the late nineteenth and early twentieth centuries the mosquito was called *Culex mosquito*, *Culex fasciatus*, and *Stegomyia fasciata*. See Christophers, *Aedes Aegypti(L.)*, 1–7.
32. Philip S. Hench Walter Reed Yellow Fever Collection. "Carlos Juan Finlay (1833–1915)," University of Virginia Health Sciences Library, http://yellowfever.lib.virginia.edu/reed/finlay.html.
33. Mark Honigsbaum, *The Fever Trail: In Search for the Cure of Malaria* (New York: Farrar, Straus and Giroux, 2001), 194–195.
34. "Color and Quinine," *NYT*, September 6, 1883, 4.
35. Paul Russell, *Man's Mastery of Malaria* (London: Oxford University Press, 1955) 47.
36. Ibid.
37. Ibid., 55, 58.
38. Ross and Grassi bitterly contested who deserved credit for the discovery. Ross's temperament and unwillingness to recognize Grassi's contributions were part of the cause for the bad feelings between the two men. Paul Russell spoke for most entomologists when he declared "but to Ross is due, for all time, the credit of being first to place a scientific finger on mosquitoes as the agents that spread malaria from man to man." Ross was awarded the Nobel Prize for this discovery in 1902. See ibid., 63.
39. George Tindall and David Shi, *America: A Narrative History*, Brief 6th ed. (New York: W. W. Norton & Company, 2004), 764.
40. Leland Howard, *Fighting Insects: The Story of an Entomologist* (New York: 1933), 118.
41. Ibid., 118–119.
42. Ibid. The literature on yellow fever and the Reed Commission has grown tremendously in recent years. Margaret Humphreys, Molly Crosby, James V. Writer, John Pierce, and Alfred Bollet offered important reassessments of the significance of Finlay's work, Walter Reed, and the Reed Commission. For Americans at the beginning of the twentieth century, the Reed Commission became synonymous with the triumph of modern medicine against an ancient foe. William Gorgas's implementation of the Commission's findings became the model for the United States Public Health Services anti-mosquito campaigns. See Humphreys, *Malaria*; Molly Caldwell Crosby, *The American Plague: The Untold Story of Yellow Fever, the Epidemic That Shaped Our History* (Berkeley: University of California Press, 2006); John R. Pierce and James V. Writer, *Yellow Jack: How Yellow Fever Ravaged America and Walter Reed Discovered Its Deadly Secrets* (New York: Wiley, 2005); and E. Chaves-Carballo, "Carlos Finlay and Yellow Fever: Triumph over Adversity," *Military Medicine* 170, no. 10 (2005).
43. Walter Reed, "Letter to Leland Howard," Philip S. Hench Walter Reed Yellow Fever Collection, University of Virginia, http://etext.lib.virginia.edu/newreed/images/small/02003004.JPG.

44. "State-Wide Mosquito Campaign Is Launched," *Daytona News* (1922), 22, Florida State Board of Health Scrapbook, Lee County Mosquito Control District, Ft. Myers, Florida.
45. Linda Nash, *Inescapable Ecologies: A History of Environment, Disease, and Knowledge* (Berkeley: University of California Press, 2006).
46. Paul Mueller discovered the toxicant properties of the chemical agent known today as DDT in 1939. In November 1942 the Orlando USDA laboratory researchers began their experiments with DDT. See Patterson, *The Mosquito Wars*, 98–99.

Chapter 1 **Waging War on the Insect Menace**

1. John Smith, *Report of the New Jersey Agricultural Experiment Station upon Mosquitoes Occurring within the State, Their Habits, Life History, Etc.* (Trenton: New Jersey Agricultural Experiment Station, 1904), 462.
2. Ibid.
3. "People of the Century: From the Ground Up," Worcester Polytechnic Institute, http://wpi.edu/News/Journal/Spring98/ground.html.
4. Spencer Miller, "Introductory Remarks," *PNJMEA* 7 (1920), 13.
5. Leland Howard, *Fighting Insects: The Story of an Entomologist* (New York: 1933), 4.
6. Arnold Mallis, *American Entomologists* (New Brunswick: Rutgers University Press, 1971), 80.
7. Howard, *Fighting Insects*, 14.
8. New York was the first state to appoint a state entomologist in 1854 (Asa Fitch). Thirteen years later Illinois followed New York's lead and named Benjamin Walsh state entomologist. Walsh was instrumental in Riley's winning the position of Missouri's entomologist in 1868. See Leland Howard, *A History of Applied Entomology* (Washington, D.C.: Smithsonian Institution Publication, 1930), 12–13; and Harlow B. Mills, *A Century of Biological Research* (New York: Arno Press, 1977), 91–92.
9. Howard, *Fighting Insects*, 14.
10. Leland Howard, *Notes on the Mosquitoes of the United States: Giving Some Account of Their Structure and Biology: With Remarks on Remedies* (Washington, D.C.: U.S. Dept. of Agriculture, Division of Entomology, 1900), 51, 55.
11. Austin W. Morrill Jr., "News and Notes," *MN* 22, no. 4 (1962), 410.
12. Robert Southey, Samuel Taylor Coleridge, and Robert Gittings, *Omniana; or, Horae Otiosiores* (Carbondale: Southern Illinois University Press, 1969), 55–56.
13. Howard, *Notes on Mosquitoes*, 52.
14. Ibid.
15. D. L. Van Dine, "Mosquito Work of the Bureau of Entomology," *American Journal of Public Health* 10, no. 2 (1920): 116.
16. Miller, "Introductory Remarks," 14.
17. Howard, *Fighting Insects*, 119.
18. Leland Howard, *Mosquitoes: How They Live; How They Carry Disease; How They Are Classified; How They May Be Destroyed* (New York: McClure Phillips & Co., 1901), vi, vii.
19. Miller, "Introductory Remarks," 14.
20. Leland Howard, "Efforts at Mosquito Control in Different Parts of the World," *PNJMEA* 7 (1920), 16.
21. Howard Markel, *Quarantine! East European Jewish Immigrants and the New York City Epidemics of 1892* (Baltimore: Johns Hopkins University Press, 1997), 73.

22. "Dr. Alvah H. Doty Dies in 80th Year," *NYT,* May 28, 1934, 19.
23. "Malarial Fever Prevalent Here," *NYT,* August 1, 1901, 12.
24. "War on Mosquitoes Begins," *NYT,* August 3, 1901, 12.
25. "A War of Elaborate Dimensions," *NYT,* August 4, 1907, 8.
26. "The War on Mosquitoes," *NYT,* August 7, 1901, 12.
27. The town of Lima derived its name from Lima, Peru. Local historians believe that the name was chosen because of the boxes of quinine that were imported from Peru to ward off the effects of malaria, which was endemic in the 1830s throughout northwestern Ohio. See http://www.cityhall.lima.oh.us/dept/local/history.asp.
28. "Mosquito Pest Abating," *NYT,* August 18, 1901, 13.
29. Howard, *Mosquitoes,* 180–181.
30. Quoted in George Bradley, "A Review of Malaria Control and Eradication in the United States," *MN* 26, no. 4 (1966): 463.
31. "Mosquito Hunting and the Goose Woman," *Hartford Courant,* August 9, 1901, 10.
32. "Dr. Doty's Mosquito Crusade Bears Fruit," *NYT,* August 10, 1901, 12.
33. Ibid.
34. "Mosquitoes Defy Science," *Los Angeles Times,* June 12, 1901, 15.
35. "Mosquitoes as Firebugs," *NYT,* July 22, 1901, 1.
36. Howard (1909) quoted in Mallis, *American Entomologists,* 83.
37. Judy Hansen, "The History of the Cape May Mosquito Control Commission" (Cape May County Mosquito Control Commission, 1999), 1.
38. John Smith, "Mosquitoes," *Entomological Notes* (September 1896): 204–205.
39. Quoted in Howard, *Mosquitoes,* 83.
40. John Smith, "Report of the Entomological Department of the New Jersey Agricultural College Experiment Station for the Year 1901" (Trenton, N.J.: John Murphy Publishing Co., 1902), 587.
41. Thomas J. Headlee, *The Mosquitoes of New Jersey and Their Control* (New Brunswick: Rutgers University Press, 1945), 264.
42. Smith, "Report for the Year 1901," 581.
43. Ibid., 580–581.
44. William Edgar Sackett, *Modern Battles of Trenton: From Werts to Wilson* (New York: Neale Publishing Company, 1914), 90–91.
45. Ibid., 92–93.
46. Ibid.
47. Ransom E. Noble, *New Jersey Progressivism before Wilson* (Princeton: Princeton University Press, 1946), 19.
48. Smith, "Report for the Year 1901," 587.
49. Smith, *Report upon Mosquitoes,* 1.
50. Ibid., 2.
51. Ibid., 3.
52. Ibid., 4.
53. Ibid., 6.
54. Smith, letter to Turner Brakeley, March 9, 1903, in JBSs.
55. Smith, letter to Leland Howard, April 2, 1903, in JBSs.
56. Ibid.
57. Smith, *Report upon Mosquitoes,* 9.
58. Smith, letter to Hermann Brehme, March 6, 1903, in JBSs.
59. Smith, letter to Otto Seifert, April 29, 1903, in JBSs.
60. Smith, letter to James Fletcher, May 4, 1903, in JBSs.
61. Dyar, letter to John B. Smith, May 15, 1903, in JBSr.

62. Harrison G. Dyar, "Annual Address of the President," *Proceedings of the Entomological Society of Washington* 5, (1903): 170, 171, 172.
63. Dyar, letter to John B. Smith, April 3, 1903, in JBSr.
64. Smith, letter to Harrison Dyar, May 16, 1903, in JBSs.
65. Miller, telegram to John B. Smith, May 11, 1903, in JBSr.
66. Miller, letter to John B. Smith, April 9, 1903, in JBSr.
67. "Mosquito Fighters in Peril," *NYT*, July 10, 1903, 5.
68. Miller, letter to John B. Smith, April 9, 1903, in JBSr.
69. Ibid.
70. Harold I. Eaton, "Atlantic County Mosquito Extermination Commission," *PNJMEA* 1 (1914), 22.
71. Smith, letter to W. W. Heberton, May 12, 1903, in JBSs.
72. Ann Vileisis, *Discovering the Unknown Landscape: A History of America's Wetlands* (Washington, D.C.: Island Press, 1997), 113.
73. Ralph Hunt, "The Anti-Mosquito Movement," *PNJMEA* 2 (1915), 11.
74. Smith, Report upon Mosquitoes, 375–379.
75. Ibid., 370.
76. Headlee, *Mosquitoes of New Jersey*, 6.
77. Smith, letter to Henry Mitchell, March 4, 1904, in JBSs.
78. Smith, *Report upon Mosquitoes*, 144.
79. Spencer Miller, "Address," *PNJMEA* 12 (1925), 18–19.
80. Smith, letter to Henry Mitchell, February 25, 1904, in JBSs.
81. Mitchell, letter to John B. Smith, February 29, 1904, in JBSr.
82. Smith, letter to Henry Mitchell, March 4, 1904, in JBSs.
83. AFI, "Smashing a Jersey Mosquito," http://www.afi.com/members/catalog/AbbrView.aspx?s=1&Movie=41653.
84. Smith, letter to Governor Franklin Murphy's secretary, February 20, 1904, in JBSs.
85. Smith, letter to Horatio Parker, February 1, 1904, in JBSs.
86. Smith, letter to Edmund Wakelee, February 26, 1904, in JBSs.
87. Smith, letter to Eugene Winship, March 10, 1904, in JBSs.
88. Smith, letter to Benjamin Morris, March 7, 1904, in JBSs.
89. Smith, letter to Governor Franklin Murphy, March 14, 1904, in JBSs.
90. Smith, letter to Byron Cummings, May 14, 1903, in JBSs.
91. Quoted in Smith, *Report upon Mosquitoes*, 442.
92. Quoted in ibid., 417.

Chapter 2 **The Garden State Takes the Lead**

1. Quayle, letter to John B. Smith, April 18, 1905, in JBSr.
2. Howard, letter to John B. Smith, February 18, 1905, in JBSr.
3. Davis, letter to John B. Smith, February 19, 1905, in JBSr.
4. Van Dine, letter to Leland Howard (John B. Smith), January 31, 1905, in JBSr.
5. H. F Gray, "The Organization and Development of Mosquito Control Work in California," *PFAMA* 19 (1948), 17.
6. C. W. Woodworth, "Mosquito Control," 1905, 1. Archive of the New Jersey Mosquito Control Association, Headlee Laboratory, New Brunswick, New Jersey.
7. University of California History Digital Archives, "Entomology and Parasitology," http://sunsite.berkeley.edu/~ucalhist/general_history/campuses/ucb/departments_e.html#entomology.
8. Woodworth, "Mosquito Control," 2.
9. Ibid.

10. Ibid., 3.
11. H. J. Quayle, "Mosquito Control Bulletin No. 178" (Sacramento: California Agricultural Experiment Station, 1906), 5.
12. Calisphere, UC Libraries, "Henry Josef Quayle, Entomology: Riverside," http://content.cdlib.org/xtf/view?docId=hb229003gf&doc.view=frames&chunk.id=div00013&toc.depth=1&toc.id=.
13. Michael Svanevik and Shirley Burgett, "Italian Immigrant Established Tone of Burlingame Club," *Hillsborough* newsletter, Winter 2005, http://www.hillsborough.net/civica/filebank/blobdload.asp?BlobID=2630.
14. Quayle, "Mosquito Control," 5.
15. Ibid., 32.
16. Quayle, letter to John B. Smith, June 15, 1905, in JBSr.
17. Quayle, "Mosquito Control," 5–6.
18. John Smith, "Report of the Entomological Department of the New Jersey Agricultural College Experiment Station for the Year 1902" (Trenton, N.J.: New Jersey Agricultural College Experiment Station, 1903), 557.
19. John Smith, "Report of the Entomological Department of the New Jersey Agricultural College Experiment Station for the Year 1903" (Trenton, N.J.: New Jersey Agricultural College Experiment Station, 1904), 654.
20. Ibid., 654–655.
21. Quayle, "Mosquito Control," 5–6.
22. Quayle, letter to John B. Smith, April 18, 1905, in JBSr.
23. Quayle, letter to John B. Smith, June 15, 1905, in JBSr.
24. Quayle, "Mosquito Control," 31–32.
25. Woodworth, letter to John B. Smith, May 10, 1905, in JBSr.
26. Smith, letter to C. W. Woodward [Woodworth], May 17, 1905, in JBSs.
27. "Warfare on Mosquitoes," *NYT*, May 23, 1903, 10.
28. "Report on Mosquito War," *NYT*, July 16, 1903, 5.
29. "Warfare on Mosquitoes," *NYT*, 10.
30. Ibid.
31. Henry C. Weeks, "Proceedings of the First General Convention to Consider the Questions Involved in Mosquito Extermination" (New York City, December 16, 1903), 9.
32. Ibid.
33. Ibid., 4.
34. Ibid., 20.
35. Ibid., 14, 15.
36. Ibid., 15.
37. Ibid., 60, 61.
38. Henry C. Weeks, "Proceedings of the Second Annual Convention of the National Mosquito Extermination Society" (New York City, December 15–16, 1904), 1, 2.
39. Sellers, letter to John B. Smith, February 13, 1905, in JBSr.
40. Smith, letter to Hermann Brehme, March 9, 1905, in JBSs.
41. Smith, letter to Benjamin Howell, March 15, 1905, in JBSs.
42. Cromwell Childe, "Mosquito Engineering," *NYT*, May 18, 1902.
43. Smith, letter to Benjamin Howell, March 15, 1905, in JBSs.
44. Smith, letter to Henry B. Kummel, March 15, 1905, in JBSs.
45. Smith, letter to G. K. Dickinson, March 15, 1905, in JBSs.
46. Smith, letter to Godfrey Mattheus, March 15, 1905, in JBSs.
47. Smith, letter to Edward Duffield, March 25, 1905, in JBSs.
48. Smith, letter to Hermann Brehme, March 30, 1905, in JBSs.

49. Smith, letter to Henry Clay Weeks, April 7, 1905, in JBSs.
50. Smith, letter to Herman Brehme, May 29, 1905, in JBSs.
51. Smith, letter to Alvah Doty, June 3, 1905, in JBSs.
52. Doty, telegram to John Smith, June 6, 1905, in JBSr.
53. Smith, letter to Porter Felt, June 13, 1905, in JBSs.
54. Smith, letter to A. N. Brown, May 29, 1905, in JBSs.
55. Smith, letter to Governor of the State of Delaware, June 19, 1905, in JBSs.
56. Smith, letter to Alvah Doty, June 19, 1905, in JBSs.
57. Cromwell, letter to John B. Smith, June 19, 1905, in JBSr.
58. Smith, letter to Porter Felt, June 21, 1905, in JBSs.
59. J. C. Bayles, "Ithaca's Formidable Health Problem," *NYT*, March 12, 1903.
60. Robert Matheson, "A Brief Account of the Malaria Epidemic at Ithaca (1904–1907)," *PNJMEA* 28 (1941): 163.
61. "War on Mosquito Pests," *Washington Post*, February 23, 1905, 4.
62. Matheson, "Malaria Epidemic at Ithaca," 163.
63. Ibid.
64. Jo Ann Carrigan, *The Saffron Scourge* (University of Southwestern Louisiana, 1994), 168.
65. Quitman Kohnke, "The Mosquito Question," *Second Annual Convention of the National Mosquito Extermination Society* (New York City, December 15–16, 1904), 77–102.
66. Carrigan, *Saffron Scourge*, 171–172, 178.
67. Ibid., 180.
68. Ibid., 195.
69. This discussion of the origin of mosquito control in Florida is drawn from Patterson, *The Mosquito Wars* (Gainesville: University Press of Florida, 2004), 19–26.
70. Hiram Byrd, "Mosquitoes of Florida" (Florida State Board of Health, 1905), 16.
71. Louisiana Division New Orleans Public Library, "Yellow Fever Deaths in New Orleans, 1817–1905," http://nutrias.org/facts/feverdeaths.htm.
72. Smith, letter to Paul Cravath, April 14, 1905, in JBSs.
73. Cravath, letter to John B. Smith, April 17, 1905, in JBSr.
74. Matheson, letter to John B. Smith, April 14, 1905, in JBSr.
75. Weeks, letter to John B. Smith, May 30, 1905, in JBSr.
76. Henry C. Weeks, "The American Mosquito Extermination Society," *Entomological News* (October 1905): 278, 279.
77. John B. Smith, "Letter to the Editors," *Entomological News* (November 1905): 306.
78. Ibid., 307.
79. Ibid.
80. Smith, letter to Spencer Miller, December 11, 1905, in JBSs.
81. Smith, letter to Conference Committee on Mosquitoes, January 16, 1906, in JBSs.
82. Patricia F. Colrick, *Spring Lake, New Jersey*, Images of America series (Portsmouth, N.H.: Arcadia, 1998), 22.
83. Smith, letter to Conference Committee on Mosquitoes, January 16, 1906, in JBSs.
84. Henry C. Weeks, "Letter to the Editors," *Entomological News* (January 1906): 15.
85. Ibid., 17–18.
86. Ibid., 21.
87. John B. Smith, "Letter to the Editors," *Entomological News* (February 1906): 69.
88. Weeks, letter to John B. Smith, January 26, 1905, in JBSr.

89. Meske, letter to John B. Smith, February 21, 1905, in JBSr.
90. Weeks died four years later in 1910. His obituary described him as "Friend of Shade Trees and Enemy of Mosquito and Telephone Companies." Weeks had spent the last years of his life fighting against "what he considered the unwarranted encroachments of the New York and New Jersey Telephone Company. The law got the best of him in the end in that fight." See "Henry Clay Weeks Dead," *NYT*, July 2, 1910, 7.
91. McFarland, letter to John B. Smith, March 26, 1906, in JBSr.
92. McFarland, letter to John B. Smith, March 2, 1905, in JBSr.
93. American Chemical Society Louisiana Section, "From Our Archives, Part II," http://membership.acs.org/l/louisiana/La100_2.htm.
94. "Shreveport in Alexandria," *Shreveport Times*, February 13, 1906.
95. Shield, letter to O. H. Brown, February 23, 1906, in JBSr.
96. Meske, letter to John B. Smith, February 14, 1906, in JBSr.
97. Smith, letter to T. N. Gray, April 6, 1906, in JBSs.
98. Smith, letter to Louis Richards, April 6, 1906, in JBSs.
99. Ibid.
100. Smith, letter to T. N. Gray, April 16, 1906, in JBSs.
101. Smith, letter to Alvah Doty, April 27, 1905, in JBSs.
102. Smith, letter to Charles Davenport, March 30, 1907, in JBSs.
103. Smith, letter to Hermann Brehme, April 23, 1906, in JBSs.
104. Smith, letter to Alvah Doty, April 27, 1906, in JBSs.
105. Smith, letter to C. C. Abbot, December 24, 1908, in JBSs.

Chapter 3 **A Continental Crusade**

1. Roger W. Lotchin, "Introduction," In *The San Francisco Calamity by Earthquake and Fire*, by Charles Morris (Urbana: University of Illinois Press, 2002), xiv. The literature on the San Francisco earthquake is extensive. Philip Fradkin argues that ineptitude in the days following the earthquake greatly magnified the damage. Developers used the breakdown of the city's municipal water service as an argument for damming the Hetch Hetchy Valley, provoking "the first major conservation battle in the twentieth century." Philip Fradkin, *The Great Earthquake and Firestorms of 1906: How San Francisco Nearly Destroyed Itself* (Berkeley: University of California Press, 2006), 5. For recent work on the earthquake and fire, see Paul H. Jeffers, *Disaster by the Bay: The Great San Francisco Earthquake and Fire of 1906* (Guilford, Conn.: Lyons Press, 2003) and Simon Winchester, *A Crack in the Edge of the World: America and the Great California Earthquake* (New York: HarperCollins, 2005).
2. H. J. Quayle, "Mosquito Control Bulletin No. 178" (Sacramento: California Agricultural Experiment Station, 1906), 33.
3. William B. Herms, "Malaria Control," *California State Board of Health Monthly Bulletin* (November 1920), 75.
4. "Notes," *Second Annual Conference of Mosquito Abatement District Officials in California* (Berkeley, California, 1931), 19.
5. William B. Herms, "Malaria Bearing Mosquitoes Live Because of Man's Own Carelessness," reprint from *Nation's Health*, August 1926: 1.
6. Smith, letter to William Herms, April 24, 1907, in JBSs.
7. William B. Herms, "Looking Back Half a Century for Guidance in Planning and Conducting Mosquito Control Operations," *PCMCA* 17 (1949), 90.
8. William C. Reeves, "The Life and Achievements of Professor William Brodbeck Herms," *PCMCA* 62 (1994), 2.

9. Herms, "Looking Back Half a Century," 90.
10. Ibid.
11. William B. Herms, "Anopheline Mosquito Investigations in California" (paper presented at the IV International Congress of Entomology, Ithaca, N.Y., 1929), 709.
12. Ibid.
13. William B. Herms, "Achievements in Malaria Control," *Transactions of the Commonwealth Club of California* 21 (1926–1927): 249.
14. H. F Gray and Russell Fontaine, "A History of Malaria in California," PCMCA 25 (1957), 19. John Work led the 1833 expedition. His diary chronicles malaria's effect on the trappers and the native population. See Linda Nash, *Inescapable Ecologies: A History of Environment, Disease, and Knowledge* (Berkeley: University of California, 2006), 20–23.
15. Quoted in Gray and Fontaine, "History of Malaria," 23.
16. Ibid., 20. See Sherburne F. Cook, "The Epidemic of 1830–1833 in California and Oregon," *University of California Publications in American Archaeology and Ethnology* 43 (1955) and Woodrow Borah and Sherburne F. Cook, *Essays in Population History*, 3 vols. (Berkeley: University of California Press, 1976).
17. "A Growing Evil in the Mines—Malaria," *NYT*, November 16, 1858, 2.
18. "Malaria in California," *Chicago Daily Tribune*, July 19, 1885, 10.
19. Herms, "Achievements in Malaria Control," 249.
20. William Herms, "Malaria on the Farm—Its Cause and Control," *University of California Journal of Agriculture* 2, no. 6 (1915): 213.
21. Herms, "Achievements in Malaria Control," 249.
22. Gray and Fontaine, "History of Malaria in California," 27.
23. Ibid., 32.
24. William Herms and Harold Gray, *Mosquito Control: Practical Methods for Abatement of Disease Vectors and Pests* (New York: The Commonwealth Fund, 1944), 136–137.
25. William Herms, *Malaria: Cause and Control* (New York: Macmillan, 1913), 89.
26. Harry E. Butler, "Citrus Colony at Penryn, Placer County, California" (California State Library, 1924), 1.
27. Ibid., 2.
28. Nash, *Inescapable Ecologies*, 109. Nash observes that "the spread of irrigated agriculture" was "especially critical to the increase in malaria."
29. Quoted in Herms, "Malaria Control," 76.
30. Gray and Fontaine, "History of Malaria in California," 26. Concern about the role of Chinese immigrants in spreading malaria surfaced in the 1860s. In the Third Biennial Report of the California State Board of Health (1874–1875), Thomas Logan disputed the claim. Logan maintained that Negroes and "the copper colored races, which include the Chinese and Indians" had a higher than average mortality to malaria. See ibid., 28–29.
31. William Herms, "The Malaria Problem," *Transactions of the Commonwealth Club of California* 11, no. 3 (March 1916): 15.
32. Stanley Bailey, "Some Ecological Aspects of Malaria in California," PCMCA 40 (1972), 90–91.
33. Herms, "Malaria Control," 77–78. Between 1900 and 1910, the California State Board of Health estimated malaria cost the state 2.8 million dollars annually. See Nash, *Inescapable Ecologies*, 96.
34. Gray and Fontaine, "History of Malaria," 34.
35. William Snow, "Malaria the Minotaur of California," *California State Board of Health Monthly Bulletin* (December 1909): 109.

36. Gray and Fontaine, "History of Malaria," 33.
37. William Snow, "Biennial Report California Department of Public Health Fiscal Years from July 1, 1908, to June 30, 1910" (Sacramento: 1910), 24.
38. Snow, "Malaria the Minotaur," 112.
39. Herms, "Looking Back Half a Century," 90.
40. "Notes," 17–18.
41. Reeves, "Life of William Brodbeck Herms," 2.
42. "Notes," 19–20.
43. William Herms, "Anti-Mosquito Organization in California," *California State Board of Health Monthly Bulletin* (1910): 313.
44. J. Albert Nyhen, "Mosquito Work in Brookline, Massachusetts," *PNJMEA* 8 (1922): 136.
45. Ibid., 136–137.
46. Ibid., 137.
47. Herms, "Looking Back Half a Century," 90.
48. Malcolm Watson, *The Prevention of Malaria in the Federated Malay States* (London: John Murray, 1921), 7–9.
49. J. W. Field and J. A. Reid, "Malaria Control in Malaya," *Journal of Tropical Medicine* (February 1956): 23.
50. Ibid., 26.
51. "Notes," 18.
52. Herms, *Malaria: Cause and Control*, 87.
53. William Herms, "Anti-Mosquito Organization in California." *California State Board of Health Monthly Bulletin* (November 1910): 314.
54. Herms, "Achievements in Malaria Control," 249, 250.
55. Reeves, "William Brodbeck Herms," 2.
56. Herms, *Malaria: Cause and Control*, 114.
57. Gray and Fontaine, "Malaria in California," 17.
58. Herms, "Anti-Mosquito Organization," 317.
59. William Herms, "The Malaria Problem." *Transactions of the Commonwealth Club of California*. 11, no. 3 (March 1916): 15.
60. "The Mosquito," *Chicago Daily Tribune*, April 25, 1906, 8.
61. Smith, letter to Robert. F. Engle, April 30, 1907, in JBSs.
62. William Edgar Darnall, "The Present Status of Mosquito Control in New Jersey," *PNJMEA* 4 (1917), 8.
63. Smith, letter to Committee on Appropriations, February 8, 1909, in JBSs.
64. "A War of Elaborate Dimensions Is Being Waged against the King of Pests in New York's Suburbs: Jersey Mosquito Has Seen His Best Days," *NYT*, August 4, 1907, 8.
65. "Alas! Poor Culex," *NYT*, November 9, 1906, 7.
66. "A War of Elaborate Dimensions."
67. Evelyn Groesbeeck Mitchell and James William Dupree, *Mosquito Life, the Habits and Life Cycles of the Known Mosquitoes of the United States; Methods for Their Control: And Keys for Easy Identification of the Species in Their Various Stages* (New York and London: G. P. Putnam's Sons, 1906), xviii.
68. Smith, letter to Evelyn Mitchell, December 10, 1907, in JBSs.
69. Smith, letter to Edward Voorhees, January 4, 1910, in JBSs.
70. Smith, letter to John Dobbins, January 8, 1910, in JBSs.
71. Smith, letter to John Dobbins, January 5, 1910, in JBSs.
72. Ralph Hunt, "The Anti-Mosquito Movement," *PNJMEA* 2 (1915): 12–13.
73. Smith, letter to Charles A. Bloomfield, February 18, 1910, in JBSs.
74. Smith, letter to Leo Schwartz, July 3, 1911, in JBSs.

75. Smith, letter to A. C. Benedict, August 12, 1911, in JBSs.
76. Smith, letter to M. N. Baker, August 12, 1911, in JBSs.
77. Smith, letter to Dr. B Van Doren Hedges, March 8, 1911, in JBSs.
78. Smith, letter to Robert. F. Engle, September 1, 1911, in JBSs.
79. Smith, letter to Susanne Strong, September 22, 1911, in JBSs.
80. Strong, letter to John B. Smith, September 1911, in JBSr.
81. Smith, letter to Turner Brakeley, October 2, 1911, in JBSs.
82. Smith, letter to Thomas Symons, September 6, 1911, in JBSs.
83. Smith, letter to William Delaney, September 25, 1911, in JBSs.
84. "Gov. Wilson and the Mosquito," *Washington Post*, October 11, 1911, 6.
85. Ibid.
86. Dobbins, letter to John Smith, November 21, 1911, in JBSr.
87. Smith, letter to John Dobbins, November 17, 1911, in JBSs.
88. Brehme, letter to John B. Smith, December 4, 1911, in JBSr.
89. Smith, letter to Hermann Brehme, January 18, 1912, in JBSs.
90. Brehme, letter to John B. Smith, December 4, 1911, in JBSr.
91. Smith, letter to John Dobbins, December 12, 1911, in JBSs.
92. Smith, letter to Edward Loomis, January 5, 1912, in JBSs.
93. Spencer Miller, "Lessons Learned from the Campaign of 1926," *PNJMEA* 14 (1927), 7.
94. Smith, letter to North Jersey Mosquito Extermination Society, January 15, 1912, in JBSs.
95. Brehme, letter to John Smith, January 16, 1912, in JBSr.
96. Smith, letter to A.C. Benedict, January 18, 1912, in JBSs.
97. Smith, letter to Franklin Dye, February 7, 1912, in JBSs.
98. Smith, letter to Turner Brakeley, March 4, 1912, in JBSs.
99. Smith, letter to Louis Richards, March 4, 1912, in JBSs.

Chapter 4 **Public Health, Race, and Mosquitoes**

1. John Solomon Otto, *The Final Frontiers, 1880–1930: Settling the Southern Bottomlands* (Westport, Conn.: Greenwood Press, 1999), 67.
2. Thomas D. Mulhern, "Thomas Headlee," *MN* 3, no. 2 (1945), 42.
3. Quoted in Thomas J. Headlee, ed., "Report of the Entomological Department of the New Jersey Agricultural College Experiment Station," (Trenton, N.J.: New Jersey Agricultural Experiment Station, 1913), 741.
4. Commissions formed in Atlantic, Essex, Hudson, Passaic, and Union counties in 1912; Ocean county in 1913; Bergen, Monmouth, Camden, and Middlesex in 1914; and Cape May in 1915.
5. Quoted in Headlee, "Report of the N.J. Agricultural College Experiment Station," 741.
6. Robert B. Gray, "Thirty-two Years of Progress 1912–1944," *PNJMEA* 31 (1944), 188.
7. Spencer Miller, "Anti-Mosquito Work in New Jersey: Methods Employed and Results Obtained," *PNJMEA* 1 (1914), 10.
8. William Delaney, "Address," *PNJMEA* 1 (1914), 16–17.
9. Thomas J. Headlee, "Anti-Mosquito Work of the New Jersey Experiment Station," *PNJMEA* 1 (1914), 36
10. Newton A. K. Bugbee, "The Attitude and Part of the State Relative to the Completion of the Drainage of the Salt Marshes," *PNJMEA* 6 (1919), 138.
11. Thomas J. Headlee, "Report on Mosquito Work for 1915" (New Brunswick: New Jersey Agricultural Experiment Station, 1916), 364.

12. "Plans Bitter Fight on Jersey Mosquito," *NYT*, March 15, 1925, RE2.
13. H. F. Gray, "Malaria Control in California," *American Journal of Public Health* (June 1912): 453.
14. William B. Herms, *Malaria: Cause and Control* (New York: Macmillan, 1913), 146.
15. Ibid., 146–150.
16. Earl W. Mortenson, "A Historical Review of Mosquito Prevention in California: Part I (1904–1946)," *PCMVCA* 53 (1985), 42.
17. Herms, *Malaria*, 96.
18. Ibid., 102.
19. Ibid., 3.
20. Ibid., vii.
21. Ibid., 139.
22. Ibid., 119.
23. Gray, "Malaria Control in California," 455.
24. Herms, *Malaria*, 119.
25. Ibid., 127.
26. Ibid., 128.
27. Ibid., 147, 150.
28. Mary L. McLendon, "The Political Rights of Women," *Atlanta Constitution*, July 13, 1913, A10.
29. H. F. Gray and Russell Fontaine, "A History of Malaria in California," *PCMCA* 25 (1957), 35.
30. Harold E. Woodworth, "Report on Mosquito Control in the Vicinity of San Mateo 1904–1915" (Berkeley, California: Agricultural Experiment Station, College of Agriculture, University of California, 1915), 3.
31. Ibid.
32. Quoted in ibid., 5.
33. Ibid., 6.
34. Ibid.
35. Ibid., 7.
36. William B. Herms, "A State-Wide Malaria-Mosquito Survey of California," *Journal of Economic Entomology* 10, no. 3 (1917): 361.
37. Bureau of Vector Control, "Three Cities Mad Scene of Anti-Mosquito Pest Movement," *Mosquito Buzz*, February 15, 1948, 2.
38. Gray and Fontaine, "History of Malaria," 35.
39. C. C. Adams, "Mosquito Control Work in Nassau County, L.I.N.Y.," *PNJMEA* 4 (1917), 176.
40. Eugene Winship, "Mosquito Extermination in Greater New York: 1900–1916," *PNJMEA* 4 (1917), 164.
41. "Mosquito Bill Is Explained," *Hartford Courant*, March 25, 1915, 10.
42. W. E. Britton, "Mosquito Control Work in Connecticut in 1916," *PNJMEA* 4 (1917), 186–187.
43. Russell W. Gies, "The Financial Side of Mosquito Control," *PNJMEA* 3 (1916), 34.
44. Herman Hornig, "Anti-Mosquito Work in Philadelphia," *PNJMEA* 1 (1914), 73, 74.
45. W. A. Evans, "Mosquito Fight Begun in Chicago Spreading Afar," *Chicago Daily Tribune*, July 16, 1914, 13.
46. W. A. Evans, "Mosquito Fight in Ravinia Wins," *Chicago Daily Tribune*, July 28, 1914, 5.
47. "Winnetka Opens Anti-Mosquito Campaign Today," *Chicago Daily Tribune*, April 15, 1916, 13.
48. Herms, "State-Wide Malaria-Mosquito Survey," 365.

49. William B. Herms, "Occurrence of Malaria and Anopheline Mosquitoes in Northern California," *Public Health Reports*, Reprint 441. (Washington, D.C.: U.S. Treasury Department, 1919), 4.
50. William B. Herms, "Progress Report on State-Wide Mosquito Survey," *California State Board of Health Report* 12, no. 4 (1916): 192.
51. William B. Herms, "Looking Back Half a Century for Guidance in Planning and Conducting Mosquito Control Operations," *PCMCA* 17 (1949), 91.
52. Ibid.
53. William B. Herms, "The Malaria-Mosquito Survey of the State," *Journal of Agriculture* 4 (1916): 42.
54. Herms, "Progress Report on State-Wide Mosquito Survey," 192, 193.
55. Bess Furman and Ralph C. Williams, *A Profile of the United States Public Health Service, 1798–1948* (Washington, D.C.: U.S. Government Printing Office, National Institutes of Health, 1973), 286–287.
56. Ralph C. Williams, *The United States Public Health Service: 1798–1950* (Washington, D.C.: Commissioned Officers Association, 1951), 295.
57. Robert D. Leigh, *Federal Health Administration in the United States* (New York: Harper and Brothers, 1927), 397.
58. Williams, *U.S. Public Health Service*, 295.
59. W. G. Stromquist, "Engineers and Malaria Control" *Transactions of the Second Annual Antimalaria Conference of Sanitary Engineers and others Engaged in Malaria Field Investigations and Mosquito Control* (Public Health Bulletin 115) (Washington, D.C.: Government Printing Office, 1921), 89.
60. R. H. von Ezdorf, "Demonstration of Malaria Control," *Public Health Reports* 31 (1916): 616.
61. Ibid., 618.
62. Ibid., 620.
63. Ibid., 622–623.
64. Quoted in ibid., 624–625.
65. The demonstration projects included Electric Mills, Mississippi (1914); Lanett, Kaulton, and Holt, Alabama (1915); Crystal City, Missouri (1915); Emporia and Wilson, Virginia (1916); and Crossett and Lake Village, Arkansas (1916).
66. Quoted in E. Richard Brown, *Rockefeller Medicine Men* (Berkeley: University of California Press, 1979), 169.
67. Williams, *U.S. Public Health Service*, 298. Also see Darwin H. Stapleton, "Lessons of History? Anti-Malaria Strategies of the International Health Board and the Rockefeller Foundation from the 1920s to the Era of DDT," *Public Health Reports* 119 (March–April 2004) and Leo Slater, *War and Disease: Biomedical Research on Malaria in the Twentieth Century* (New Brunswick: Rutgers University Press, 2009).
68. R. C. Derivaux, H. A. Taylor, and T. D. Haas, "Malaria Control: A Report of Demonstration Studies Conducted in Urban and Rural Sections," Public Health Bulletin no. 88., ed. United States Public Health Service (Washington, D.C.: Government Printing Office, 1917), 55.
69. Ibid., 14.
70. Ibid., 35.
71. A. W. Fuchs, "Malaria Statistics," *Transactions of the Second Annual Antimalaria Conference of Sanitary Engineers and others Engaged in Malaria Field Investigations and Mosquito Control (Public Health Bulletin 115)* (Washington, D.C.: Government Printing Office, 1920), 177–178.
72. C. E. Buck, "Surveys and Maps," *Transactions of the Second Annual Antimalaria Conference*, 183.

73. Bruce Mayne, "The Part Biology Plays," *Transactions of the Second Annual Antimalaria Conference*, 120.
74. L. M. Fisher, "The Organization of Malaria-Control Division in State Board of Health," *Transactions of the Second Annual Antimalaria Conference*, 69.
75. James Hayne, "Malaria," *Transactions of the Nineteenth Annual Conference of State and Territorial Health Officers with the United States Public Health Service* (Washington, D.C.: Public Health Bulletin 111, 1922), 64, 65.
76. James Hayne, "Malaria-Control Investigations Discussion," *Transactions of the Second Annual Antimalaria Conference*, 35.
77. W. H. De Mott, "Mosquito Organization in Nassau County, Long Island, N.Y," *Transactions of the Second Annual Antimalaria Conference*, 21.
78. "Man's Great Enemy . . . the Mosquito," *Florida Health Notes*, March 1953, 56.
79. Elisabeth Beck, *Public Health Pest Control Manual: Applicator Training Manual*, revised by Thomas Loyless, Florida Department of Agriculture and Consumer Services, Tallahassee, 1999, 18.
80. Williams, *U.S. Public Health Service*, 299–300.
81. Ibid., 561.
82. Leigh, *Federal Health Administration in the United States*, 116.
83. Williams, *U.S. Public Health Service*, 584–585.
84. Ibid., 296.
85. "Malaria Control Work," *Florida Health Notes* 13 no. 3 (November 1918), 80. See also Patterson, *The Mosquito Wars: A History of Mosquito Control in Florida* (Gainesville: University Press of Florida, 2004), 34–36.
86. "Malaria Control Work," *Florida Health Notes* 13, no. 3 (November 1918): 80.
87. William Williams, "Progress in Mosquito Work in Bergen County," *PNJMEA* 5 (1918), 12.
88. Joseph Albert Augustin LePrince, "Mosquito Control About Cantonments and Shipyards," *PNJMEA* 6 (1919), 14.
89. "Negro Soldier Is Used for Mosquito Bait," *Arkansas Democrat*, July 15, 1918.
90. William B. Herms, "The Mosquito Survey," *California State Board of Health Report* 13, no. 6 (1917): 267.
91. Quoted in William B. Herms, "Malaria Drainage Operations at the Port of Embarkation, Newport News, Virginia," *The Military Surgeon*, no. July (1920): 1.
92. Mark Boyd, "A Historical Sketch of the Prevalence of Malaria in North America," *American Journal of Tropical Medicine* 21 (1941): 226.
93. H. R. Carter, "The Malaria Problem in the South," *PNJMEA* 4 (1917), 88.
94. W. J. Schoene, "Mosquito Situation in Virginia," *PNJMEA* 4 (1917), 193.
95. Herms, "Malaria Drainage Operations," 2.
96. Ibid., 11.
97. Ibid., 1.
98. L. D. Fricks, "Malaria," *Transactions of the Eighteenth Annual Conference of State and Territorial Health Officers with the United States Public Heath Service* (Washington, D.C.: Public Health Bulletin 111, April 1921), 90.
99. William B. Herms, "What Shall We Do with Our Information Concerning Malaria in California" (Sacramento: California State Board of Health, California State Printing Office, 1920), 12.

Chapter 5 **Widening the Campaign**

1. D. H. Lawrence, "The Mosquito," *Bookman* (July 1921), 430–431.
2. William Faulkner, *William Faulkner: Novels 1926–1929* (Literary Classics of the United States, 2006), 260.

3. William B. Herms, "Occurrence of Malaria and Anopheline Mosquitoes in Northern California," *Public Health Reports*, Reprint 441. (Washington, D.C.: U.S. Treasury Department, 1919): 3.
4. William B. Herms, "What Shall We Do with Our Information Concerning Malaria in California" (Sacramento: California State Board of Health, California State Printing Office, 1920), 3.
5. H. F. Gray, "The Cost of Malaria," *Journal of the American Medical Association* 72, no. 22 (1919): 1533.
6. Ibid., 1534.
7. Louva G. Lenert and Edward Ross, "Malaria in California" (Sacramento: California State Board of Health, 1923), 13, 16.
8. Ibid., 16.
9. Linda Nash, *Inescapable Ecologies: A History of Environment, Disease, and Knowledge* (Berkeley: University of California Press, 2006), 99.
10. Gray, "The Cost of Malaria," 1534.
11. Gray, "Personal History," Archives, Greater Los Angeles Vector Control, Santa Fe Springs, California, 1954, 6.
12. Stanley B. Freeborn, "Malaria Control: A Report of Demonstration Studies at Anderson, California," *California State Board of Health Monthly Bulletin* 15, no. 9 (1920): 280.
13. Ibid., 284, 285.
14. Ibid., 287.
15. "Malaria, Last Year and This," *Anderson Valley News*, August 27, 1920, 5.
16. Herms, "Malaria Control," *California State Board of Health Monthly Bulletin* 15, no. 5 (1920), 78.
17. Nash, *Inescapable Ecologies*, 85.
18. William B. Herms, "Malaria Bearing Mosquitoes Live Because of Man's Own Carelessness," reprint from *Nation's Health*, August 1926: 1.
19. Lenert and Ross, "Malaria in California," 5–6.
20. Herms, "Progress Report on State-Wide Mosquito Survey," *California State Board of Health Report* 12, no. 4 (1916): 195. Herms fears proved unfounded. Though the number of *Anopheles freeborni* mosquitoes increased the malaria rate declined in the Sacramento Valley after 1920. See H. F. Gray and Russell Fontaine, "A History of Malaria in California," PCMCA 25 (1957), 19.
21. Lenert and Ross, "Malaria in California," 65.
22. Ibid.
23. Herms, "What Shall We Do," 8. See also Stanley Freeborn, "The Mosquitoes of California" (Ph.D. diss., Massachusetts Agricultural College, Amherst, Massachusetts, 1924).
24. Quoted in Joseph Albert Augustin LePrince and A. J. Orenstein, *Mosquito Control in Panama: The Eradication of Malaria and Yellow Fever in Cuba and Panama* (New York, London: Putnam, 1916), iv.
25. Leland Howard, "Mosquito Observations in Different Countries during the Past Three Years," *PNJMEA* 10 (1923), 14.
26. Joseph Albert Augustin LePrince, "General Plan for Mosquito Work in the Southern States and Its Apparent Results," *PNJMEA* 10 (1923), 29–30.
27. L. D. Fricks, "Service Policy and Administration Methods in Malaria-Control Investigations," *Transactions of the Second Annual Antimalaria Conference of Sanitary Engineers and others Engaged in Malaria Field Investigations and Mosquito Control*, Public Health Bulletin 115 (Washington, D.C.: United States Public Health Service, Government Printing Office, 1920) 28.

28. "Most Extensive Malarial Eradication Work in the United States Being Conducted by State Board of Health in Perry, Fla," *Sunday Times-Union* (Jacksonville: Florida State Board of Health Scrapbook, Lee County Mosquito Control District, Ft. Myers, Florida, 1920), 1. See also Gordon Patterson, *The Mosquito Wars* (Gainesville: University Press of Florida, 2004), 36–40.
29. George W. Simons, "Bureau of Engineering," *Thirty-Second Biennial Report of the State Board of Health of Florida* (Jacksonville: Florida State Board of Health, 1923), 197.
30. Ibid., 196.
31. Albert Hardy and May Pynchon, "Millstones and Milestones: Florida's Public Health from 1889" (Jacksonville: Florida State Board of Health, 1964), 33.
32. Simons, "Bureau of Engineering," 197, 199.
33. Raymond Turck, "Address to Anti-Mosquito Conference" (Daytona Beach: Florida State Board of Health Scrapbook, Lee County Mosquito Control District, Ft. Myers, Florida, 1922).
34. Grady McWhiney, *Cracker Culture: Celtic Ways in the Old South* (University: University of Alabama Press, 1988), 51; and J. S. Bradford, "Crackers," *Lippincott's Magazine of Literature, Science and Education* (November 1870), 458, 465.
35. Turck, "Address to Anti-Mosquito Conference," 31.
36. David Starr Jordan was probably the first person to note the usefulness of *Gambusia affinis* for mosquito control. Jordan made this discovery in December 1883, fewer than a hundred miles from Brewster. Jordan, later president of Stanford University, became one of the earliest advocates of "biological control" of mosquitoes. In 1900 his student, Alvin Seale, introduced *Gambusia affinis* into Texas and Hawaii and, later, in the Philippines and Taiwan. See David Starr Jordan, *The Days of a Man: Being Memories of a Naturalist, Teacher and Minor Prophet of Democracy*, vol. 1, 1851–1899 (Yonkers-on-Hudson, N.Y.: World Book Company, 1922), 289.
37. "New Jersey's Mosquito War," *NYT*, August 3, 1902, 21.
38. Ibid. Frothingham was one of the first to protest the introduction of *Gambusia affinis*. For a review of the literature on G. *affinis* and mosquito control see Henry Rupp, "Adverse Assessments of *Gambusia affinis*: An Alternate View for Mosquito Control Practitioners," *Journal of the American Mosquito Control Association* 12, no. 2 (1996). Concern about the effect of G. *affinis* on "rare indigenous species" surfaced in the 1930s (156). It was not, however, until the 1970s, that ichthyologists and ecologists questioned the effectiveness of G. *affinis* as a means of controlling mosquitoes and raised serious objections to the predacious fish's effect on nontarget species.
39. John B. Smith, "Report of the New Jersey Agricultural Experiment Station upon Mosquitoes Occurring within the State, Their Habits, Life History, Etc." (Trenton: Report of the New Jersey Agricultural Experiment Station 1904), 92.
40. Turck, "Address to Anti-Mosquito Conference," 31.
41. Ibid.
42. "Anti-Mosquito Association to Fight for Entire Elimination of Malaria Mosquito" (Florida State Board of Health Scrapbook, Lee County Mosquito Control District, Ft. Myers, Florida, 1923), 62.
43. John Pearn and Kenneth D. Winkel, "Toxinology in Australia's Colonial Era: A Chronology and Perspective of Human Envenomation in 19th Century Australia," *Toxicon* 48 (2006): 734; see also Duane J. Gubler, "Dengue and Dengue Hemorrhagic Fever: Its History and Resurgence as a Global Public Health Problem," in *Dengue and Dengue Hemorrhagic Fever*, ed. Duane J. Gubler and G. Kuno (Oxon, U.K.: CAB International, 1997), 7.

44. Howard, "Mosquito Observations in Different Countries," 24.
45. Ibid., 23.
46. George W. Simons, "The Prospects for State-Wide Mosquito Control in Florida," *PNJMEA* 10 (1923), 107.
47. "State-Wide Mosquito Campaign Is Launched," *Daytona News* (Florida State Board of Health Scrapbook, Lee County Mosquito Control District, Ft. Myers, Florida, 1922), 1.
48. Quoted in "Declare War on Mosquitoes," *St. Petersburg Times*, December 7, 1922.
49. Vida Lester MacDonell, "History of the Florida Anti-Mosquito Association," *PFAMA* 10 (1932), 3.
50. Simons, "Prospects for Mosquito Control," 103.
51. "May Request Larger State Health Levy," *St. Petersburg Times*, March 22, 1923, A1.
52. LePrince, "General Plan for Mosquito Work," 31.
53. Quoted in "Minutes of the Board Meetings of the Salt Lake City Mosquito Abatement District" (Trustees of the Salt Lake City Mosquito Abatement District, 1928).
54. Ibid.
55. D. Rees and R. W. Chamberlin, "Survey of the Mosquitoes of Salt Lake City for 1930," transcript in the Salt Lake City Mosquito Abatement District archive, 1931, 1.
56. Charles Dickens, *American Notes*, ed. John S. Whitley and Arnold Goldman (Penguin, 1985), 190.
57. Quoted in Margaret Humphreys, *Malaria: Poverty, Race, and Public Health in the United States* (Baltimore: Johns Hopkins University Press, 2001), 32.
58. Isaac D. Rawlings, *The Rise and Fall of Disease in Illinois*, 2 vols. (Springfield, Ill.: Schnepp & Barnes, 1927), 1:113.
59. Ibid., 1:72.
60. Otto McFeely, "J. Lyell Clarke," *MN* 13, no. 3 (1953): 215.
61. Leland Howard, "Letter to J. A. LePrince," http://etext.lib.virginia.edu/newreed/images/large/C0134004.JPG.
62. Rawlings, *Rise and Fall of Disease in Illinois*, 72.
63. L. O. Howard, "Mosquito Work in the Year of 1925," *PNJMEA* 13 (1926), 16.
64. Ibid., 17.
65. T.H.D. Griffitts, "Some Phases of the Salt-Marsh Mosquito Problem in the South Atlantic and Gulf States," *PNJMEA* 15 (1928), 87.
66. Ibid.
67. F. W. Harden, H. R. Hepburn, and B. J. Etheridge, "A History of Mosquitoes and Mosquito-Borne Diseases in Mississippi 1699–1965," *MN* 27, no. 1 (1967): 63.
68. Leland Howard, "What Workers with Mosquitoes Did during 1926," *PNJMEA* 14 (1927), 20.
69. *Culex salinarius*, *Anopheles crucians*, and *Anopheles atropos* were also present in the salt marshes. See Griffitts, "Some Phases of the Salt-Marsh Mosquito Problem," 89.
70. Ibid., 80, 88.
71. William Herms and Harold Gray, *Mosquito Control: Practical Methods for Abatement of Disease Vectors and Pests*. (New York: The Commonwealth Fund, 1944), 38.
72. Howard, "Mosquito Observations," 13.
73. Leland Howard, "Mosquito Work Throughout the World during 1927," *PNJMEA* 15 (1928), 14.

74. Elton J. Hansens and Harry B. Weiss, "Entomology in New Jersey" (New Brunswick: New Jersey Agricultural Experiment Station, 1953), 8.
75. D. J. Sutherland, "From the Mosquito Archives: When Were Airplanes First Used in Mosquito Control?" *New Jersey Mosquito Control Association, Inc. Newsletter*, no. 15 (3) (2003), http://www.rci.rutgers.edu/~insects/vol15no3.htm.
76. Howard, "What Workers with Mosquitoes Did during 1926," 14.
77. Ibid., 21.
78. Ibid., 22.
79. Ibid., 16.
80. Ibid.
81. Quoted in Marc E. Epstein and Pamela M. Henson, "Digging for Dyar: The Man Behind the Myth," *American Entomologist* 38, no. 3 (1992): 162.
82. Leland Howard, "The Work with Mosquitoes around the World in 1928," *PNJMEA* 16 (1929), 8.
83. William T. Forbes, "Harrison Gray Dyar: Obituary," *Entomological News* 40 (May 1929): 167.
84. Harrison G. Dyar, *The Mosquitoes of the Americans*, Carnegie Institution of Washington Publication No. 387 (Washington, D.C.: Carnegie Institution of Washington, 1928), 213.
85. Herms and Gray, *Mosquito Control*, 3, 4.
86. T.H.D. Griffitts, "A Further Report on the Salt Marsh Problem of the South Atlantic and Gulf States and Malaria Control on a County-Wide Basis," *PNJMEA* 17 (1930), 151.

Chapter 6 **Advances and Retreats during the Great Depression**

1. Thomas J. Headlee, "The Relation of Rainfall to the Seasonal 'Peak Load' of Mosquito Control," *PNJMEA* 16 (1929), 75.
2. Will Rogers, "It Is Hoped the Lady Mosquito Will Read Birth Control Literature," http://home.earthlink.net/~bobhaught/willrogersok_whatsnew/id1.html.
3. L. E. Jackson, W. H. Hobenberg, E. L. Filby, H. W. Green, and R. Gies, "Mosquito Control," *American Journal of Public Health* 16, no. 3 (1926): 258, 259, 260.
4. Judith Wrenn, "The History of Connecticut's State Mosquito Program" (Practicum, University of Connecticut, 2000), 5.
5. Leland Howard, "The Work with Mosquitoes around the World in 1929," *PNJMEA* 17 (1930), 28.
6. Jackson et al., "Mosquito Control," 258.
7. Harold I. Eaton, "Application of Mechanics to Mosquito Control," *PNJMEA* 5 (1918), 44–45.
8. Jackson et al., "Mosquito Control," 258.
9. Frank Miller, "A Summary of New Jersey Mosquito Control Accomplishments of the Past Year," *PNJMEA* 17 (1930), 51.
10. "Summary: 1932–1936," *PFAMA* 10 (1936), 1–2.
11. "Minutes," *PMCA* 1 (1930), 2.
12. H. F. Gray, "Comments," *PMCA* 3 (1932), 14–15.
13. Ibid., 16.
14. Ibid., 15.
15. Fred A. Reiley, "The Effect of Reduced Appropriations on Anti-Mosquito Work in New Jersey," *PNJMEA* 22 (1935), 162.
16. Jacob Lipman, "The Effect of the Depression upon Mosquito Control and How It May Be Overcome," *PNJMEA* 23 (1936), 54.
17. Reiley, "Effect of Reduced Appropriations," 163.

18. F. C. Bishopp, "Some Accomplishments in Mosquito Work Throughout the World During 1932," *PNJMEA* 20 (1933), 80.
19. Edward Wright, "Mosquito Control Work of the Massachusetts Reclamation Board," *PNJMEA* 21 (1934), 114.
20. Quoted in Samuel Eliot Morison, *The Oxford History of the American People* (New York: Oxford University Press, 1965), 950.
21. Quoted in E. L. Bishop, "TVA's New Deal in Health," *American Journal of Public Health* 24 (1934): 1023.
22. F. E. Gartrell, "Malaria Control Program of the Tennessee Valley Authority," *MN* 11, no. 3 (1951): 136. See also F. E. Gartrell, Joseph C. Cooney, George P. Chambers, and Ralph H. Brooks, "TVA Mosquito Control 1934–1980: Experience and Current Program Trends and Developments," *MN* 41, no. 2 (1981).
23. W. G. Stromquist, "Malaria Control in the Tennessee Valley," *Civil Engineering* 5, no. 12 (1935): 772.
24. Ibid.
25. United States Public Health Service and Tennessee Valley Authority, Health and Safety Department, *Malaria Control on Impounded Water* (Washington, D.C.: U.S. Government Printing Office, 1947), 9.
26. Stromquist, "Malaria Control," 772.
27. Gartrell et al., "Malaria Control Program of TVA," 302.
28. Stromquist, "Malaria Control," 773.
29. Ibid., 774.
30. C. E. Waller, "A Review of the Federal Civil Works Projects of the Public Health Service," *Public Health Reports* 49, no. 33 (August 17, 1934), 962–963.
31. Louis Williams Jr., "Civil Works Administration Emergency Relief Administration Malaria Control Work in the South," *American Journal of Public Health* 25, no. 1 (1935), 13.
32. Ibid., 11.
33. Louva G. Lenert, "Thirty-Fifth Annual Report of State Board of Health of Florida," Bureau of Engineering (1935), 80.
34. F. C. Bishopp, "Resume of Mosquito Work Throughout the World in 1933," *PNJMEA* 21 (1934), 37.
35. Ibid., 53–54, 56.
36. Lester Smith, "Mosquito Work of the Civil Works Administration throughout the County" *PNJMEA* 21 (1934), 70.
37. William E. Leuchtenburg, *Franklin D. Roosevelt and the New Deal: 1932–1940* (New York: Harper & Row, 1963), 122.
38. Bishopp, "Resume of Mosquito Work . . . 1933," 54–55.
39. R. V. Chamberlin and D. Rees, "Accomplishments in Mosquito Control in Utah," *PNJMEA* 22 (1935) 96, 97.
40. Lewis T. Nielsen, "Don Merrill Rees, 1901–1976: A Tribute to the Founder of the Utah Mosquito Abatement Association," *PUMAA* 50 (1997), 1.
41. William H. Behle, *History of Biology at the University of Utah (1869–200)* (Salt Lake City: University of Utah Publications and Printing Services, 2002), 112.
42. Chamberlin and Rees, "Accomplishments in Mosquito Control in Utah," 96.
43. Donald MacCreary and L. A. Stearns, "Effect of Drainage Work Accomplished by the CCC upon the Prevalence of Mosquitoes at Lewes, Delaware, during 1934," *PNJMEA* 22 (1935), 115.
44. W. S. Corkran, "Drainage Works Accomplished in Delaware by the CCC and CWA," *PNJMEA* 22 (1935), 105.
45. Ibid., 108.
46. Ibid., 108–109.

47. L. A. Stearns, "The Present Mosquito Control Situation in Delaware," *PNJMEA* 27 (1940), 106.
48. Ibid., 110.
49. Thomas D. Mulhern, "A Summary of Mosquito Control Accomplishments in New Jersey in 1935," *PNJMEA* 23, (1936), 45.
50. Fred A. Reiley, "The CCC in Mosquito Work in Southern New Jersey," *PNJMEA* 23 (1936), 130.
51. Stearns, "Present Mosquito Control Situation in Delaware," 110.
52. Ibid., 111.
53. "Warfare on Mosquitoes," *NYT*, May 23, 1903, 10.
54. C. C. Adams, "Mosquito Control Work in Nassau County, L.I.N.Y.," *PNJMEA* 4 (1917), 176–77.
55. A. C. Froeb, "Accomplishments of Mosquito Control in Suffolk County, Long Island," *PNJMEA* 23 (1936), 128.
56. Ibid.
57. "William Vogt, Former Director of Planned Parenthood, Is Dead," *NYT*, July 12, 1968, 31.
58. George Greenfield, "News of Activities with Rod and Gun," *NYT*, October 24, 1934, 30.
59. Raymond R. Camp, "Wood, Field and Stream: Reviews Loss to Farmers," *NYT*, October 22, 1937, 32.
60. "Wood, Field and Stream: Another Drainage Critic Listening in on a Catch," *NYT*, March 30, 1938, 27.
61. Eric G. Bolen, "In Memoriam: Clarence Cottam," *The Auk* 92 (1975): 120.
62. Ibid., 121.
63. Ibid., 123.
64. F. C. Bishopp, "Resume of Mosquito Work throughout the World in 1934," *PNJMEA* 22 (1935), 50.
65. Charles Forbes, "Address by the President," *PNJMEA* 22 (1935), 5.
66. Ibid., 6.
67. Thomas D. Mulhern, "Draft: Chronological Historical Notes American Mosquito Control Association, Inc." (1963, 2–3), Thomas Mulhern's Papers, Red Binder Mosquito and Vector Control Association of California Archives, Greater Los Angeles Vector Control District, Santa Fe Springs, California.
68. Ibid.
69. Ibid., 6–7.
70. Ibid.
71. William Vogt, *Thirst on the Land: A Plea for Water Conservation for the Benefit of Man and Wild Life*, circular no. 32 (New York City: National Association of Audubon Societies, 1937), 16.
72. Quoted in Vogt, *Thirst on the Land*, 17.
73. Ibid., 5.
74. Ibid., 7.
75. Ibid., 19, 15.
76. Milton H. Price, "New Developments in Mosquito Control in Rhode Island," *PNJMEA* 25 (1938), 112.
77. C. Cottam, W. S. Bourn, F. C. Bishopp, L. L. Williams, and W. Vogt, "What's Wrong with Mosquito Control," *Transactions of the North American Wildlife Conference* 3 (1938), 81, 82.
78. Ibid., 85–86.
79. Ibid., 87–88.
80. Ibid., 92.

81. Ibid.
82. Ibid., 94.
83. Ibid.
84. Ibid., 94–95.
85. Ibid., 96, 98.
86. Ibid., 97, 99.
87. "Chosen National Director of Parenthood Federation," *NYT*, May 18, 1951, 33.

Chapter 7 **Weapons of Mass Destruction**

1. Morris A. Stewart, W. T. Harrison, and H. F. Gray, "Symposium on Encephalitis: Discussion," *PCMCA* 12 (1941), 54.
2. Stanley B. Freeborn, "Military Mosquito Control in World War I," *PCMCA* 12 (1941), 61.
3. W. A. Hardenbergh and Lloyd K. Clark, "Mosquito Control in the Army Reservations in the Middle Atlantic and Northeastern States," *PNJMEA* 29 (1942), 27
4. Elizabeth Etheridge, *Sentinel for Health: A History of the Centers for Disease Control* (Berkeley: University of California Press, 1992), 2–3.
5. Darwin H. Stapleton, "Lessons of History? Anti-Malaria Strategies of the International Health Board and the Rockefeller Foundation from the 1920s to the Era of DDT," *Public Health Reports* 119 (March–April 2004): 213.
6. Quoted in ibid.
7. Margaret Humphreys, *Malaria: Poverty, Race, and Public Health in the United States* (Baltimore: Johns Hopkins University Press, 2001), 152.
8. L. L. Williams Jr., "Malaria Control in Defense Areas," *PNJMEA* 29 (1942), 36.
9. "Malaria Control Work Praised by Engineers," *Pensacola Journal* (Pensacola Project Scrapbook, Lee County Mosquito Control District, Ft. Myers, Florida), April 6, 1940.
10. S. C. Drews and W. W. Morrill Jr., "DDT for Insect Control at Army Installations in the Fourth Service Command," *Journal of Medical Entomology* 39, no. 3 (1946): 347.
11. Other students included Deane Furman, Ray Smith, Gordon Lindsay, Charles Mitchener, and George Ferguson.
12. Richard F. Peters, "Dick Peter's Oral History," 3. Mosquito and Vector Control Association of California Oral History Collection, Santa Fe Springs, California, 1994.
13. William C. Reeves, "Arbovirologist and Professor, UC Berkeley School of Public Health: An Oral History Conducted in 1990 and 1991 by Sally Smith Hughes," 21. Regional Oral History Office, Berkeley, Bancroft Library, University of California, 1993.
14. Peters, "Dick Peter's Oral History," 8.
15. Ibid., 11.
16. Richard F. Peters, "Mosquito Breeding and Control in the Vicinity of Military Zones," *PCMCA* 12 (1941), 63–64.
17. There are no accepted abbreviations for the diseases caused by infections of the WEE, SLE, and EEE viruses. A few viral diseases transmitted by mosquitoes have accepted abbreviations such as WNF for West Nile fever, caused by West Nile virus (WNV), and DHF for dengue hemorrhagic fever. In the text, WEE, SLE, and EEE refer to the pathogens that cause the encephalomyelitis. I am indebted to Bruce Eldridge for making this distinction.
18. Peters, "Dick Peter's Oral History," 11–12.
19. Reeves, "Arbovirologist and Professor," 21.

20. Ibid., 32–33.
21. Ibid., 33.
22. Ibid., 34.
23. Ibid., 35.
24. F. C. Bishopp, "Insect Problems in World War II with Special References to the Insecticide DDT," *American Journal of Public Health* 35, no. 4 (1945): 378.
25. H. H. Stage, "Mosquito Control Work in the Pacific Northwest," *PNJMEA* 25 (1938), 188–189.
26. Arthur Woody, "Mosquito Control Work in Portland, Oregon and Vicinity in 1950," *MN* 11, no. 3 (1951): 158.
27. Edward Shaw and C. M Saunders, "Karl Friedrich Meyer, Pathology; Microbiology: San Francisco," http://content.cdlib.org/xtf/view?docId=hb9k4009c7&doc.view=frames&chunk.id=div00034&toc.depth=1&toc.id=&brand=oac&query=Karl%20Frederick%20Meyer.
28. Reeves, "Arbovirologist and Professor," 40–41.
29. R. V. Chamberlin, "History of St. Louis Encephalitis," in *St. Louis Encephalitis*, ed. Thomas P. Monath (Washington, D.C.: American Public Health Association, 1980), 7.
30. Robert Matheson, "Important Mosquito Borne Diseases That Are Likely to Affect the Middle Atlantic and Northeastern States," *PNJMEA* 29 (1942), 9.
31. William C. Reeves and Marilyn M. Milby, "Strategies and Concepts for Vector Control," in *Epidemiology and Control of Mosquito-Borne Arboviruses in California, 1943–1987*, ed. William C. Reeves (Sacramento: California Mosquito and Vector Control Association, Inc., 1990), 1, 3, 5.
32. Malcolm H. Merrill, "The Mosquito as a Possible Vector of Equine Encephalomyelitis," *PCMCA* 11 (1940), 23.
33. William B. Herms, "Discussion," in *Conference of Mosquito Abatement District Officials in California* (Berkeley, California, Mosquito Abatement District, 1933), 8.
34. Reeves, "Arbovirologist and Professor," 46.
35. Ibid., 110.
36. Reeves and Milby, "Strategies and Concepts for Vector Control," 6.
37. Quoted in "Mosquitoes Declared Guilty Sleeping Sickness Carriers," *MN* 1, no. 4 (1941): 18, 20.
38. F. C. Bishopp and H. H. Stage, "A Review of Mosquito Work throughout the World in 1942," *PNJMEA* 30 (1943), 99.
39. Albert E. Cowdrey, *Fighting for Life: American Military Medicine in World War II* (New York: The Free Press, 1994), 44.
40. Ibid., 60.
41. C. B. Hutchinson, "University Welcome," *PCMCA* 17 (1949), 2.
42. Paul Russell, *Man's Mastery of Malaria* (London: Oxford University Press, 1955), 117.
43. "Willard V. King," *MN* 30, no. 2 (1970): 309.
44. Peters, "Dick Peter's Oral History," 13.
45. Hutchinson, "University Welcome," 2.
46. Austin W. Morrill Jr., "News and Notes," *MN* 16, no. 4 (1956): 312.
47. G. H. Bradley and F. Earle Lyman, "Mosquito Control Activities of the Communicable Disease Center, U.S. Public Health Service," *PNJMEA* 34 (1947), 46.
48. G. H. Bradley, "Mosquito Control Operations of the Office of Malaria Control in War Areas," *PFAMA* 17 (1946), 45.
49. Federal Security Agency, *Malaria Control in War Areas Monthly Report* (Atlanta: U.S. Public Health Service, July 1942), 2.

50. F. C. Bishopp and H. H. Stage, "A Review of Contributions to the Knowledge of Mosquitoes during 1944 in a World at War," *PNJMEA* 32 (1945), 23. See also Gordon Patterson, *The Mosquito Wars: A History of Mosquito Control in Florida* (Gainesville: University Press of Florida, 2004), 92–93, 95.
51. G. H. Bradley, "Malaria Control in War Areas in the United States in 1943 by the U.S. Public Health Service," *PNJMEA* 31 (1944), 131.
52. Stanley B. Freeborn, "The Malaria Control Program of the U.S. Public Health Service among Civilians in Extra-Military Areas," *Journal of the National Malaria Society* 3, no. 1 (1944): 21.
53. Ibid.
54. See Patterson, *The Mosquito Wars*, 95–105.
55. E. F. Knipling, "Insect Control Investigations of the Orlando, Fla., Laboratory during World War II," *Smithsonian Report for 1948 (Publication 3968)* (1949): 333.
56. Bishopp, "Insect Problems in World War II," 376.
57. In the early 1930s, Ginsburg discovered that a pyrethrum-oil mixture (New Jersey Mosquito Larvicide) was "non-injurious to water fowl, vegetation and fish, and can be applied at low cost." See Joseph Ginsburg, "Studies of Pyrethrum as a Mosquito Larvicide," *PNJMEA* 17 (1930), 57–72; and Fred .C. Bishopp, "What Workers with Mosquitoes Have Accomplished around the World in 1930 and 1931," *PNJMEA* 19 (1932), 34. Louis Williams reported in 1933 that the Virginia State Health Department had successfully employed Ginsburg's formulation against adult salt marsh mosquitoes. This was the first effective adulticide. See L. L. Williams Jr., "Mosquito Control Activities and Investigations of the United States Public Health Service," *PNJMEA* 19 (1932), 66. Lyell Clarke reported that "Ginsburg's brainchild" had made many friends for mosquito control "among lovers of birds, fish, and wild flowers. . . . Conservationists look on it with satisfaction." Lyell Clarke, "Progress of the Mosquito Control Campaign in the Des Plaines Valley Area," *PNJMEA* 23 (1936), 99.
58. Andrew Spielman and Michael D'Antonio, *Mosquito: A Natural History of Our Most Persistent and Deadly Foe* (New York: Hyperion, 2001), 135.
59. Knipling, "Insect Control Investigations," 337.
60. Humphreys, *Malaria*, 56.
61. H. H. Stage, "The Aerial Application of Insecticides for Mosquito Control," *PNJMEA* 37 (1950), 77.
62. Division of Insects Affecting Man and Animals, ed. "Report of Work Conducted under Contract with the Office of the Surgeon General, War Department, April 1942–October 1945" (Washington, D.C.: USDA Research Administration, 1945), 41.
63. Quoted in Edmund Russell, *War and Nature: Fighting Humans and Insects with Chemicals from World War I to Silent Spring* (Cambridge: Cambridge University Press, 2001), 176.
64. Joseph Ginsburg, "Progress in the Development of DDT Mosquito Larvicides," *PNJMEA* 32 (1945), 54–55.
65. Phillip Granett, "The Development of a Practical Mosquito Repellent," *PNJMEA* 27 (1940), 38.
66. C. A. Setterstrom, "Banishing Bug Bites," *MN* 6, no. 4 (1946): 187.
67. F. C. Bishopp, "Contributions of the Bureau of Entomology and Plant Quarantine of the Department of Agriculture to the National Program for Control of Malaria," *Journal of the National Malaria Society* 3, no. 1 (1944): 53.
68. Setterstrom, "Banishing Bug Bites," 186.

69. Emory C. Cushing, *History of Entomology in World War II* (Washington, D.C.: Smithsonian Institution, 1957), 38, 48–49.
70. F. C. Bishopp, "Present Position of DDT in the Control of Insects of Medical Importance," *American Journal of Public Health* 36, no. 6 (1946): 599.
71. Knipling, "Insect Control Investigations," 338.
72. Fred A. Reiley and Robert L. Vannote, "Review of Advances in Practical Methods of Mosquito Control," *PNJMEA* 25 (1938), 81.
73. Bishopp, "Contributions of the Bureau," 51–52.
74. Cushing, *History of Entomology in World War II*, 36.
75. Emory C. Cushing, "An Informal Discussion of the Work of the Bureau of Entomology and Plant Quarantine on Mosquito Control," *PNJMEA* 29 (1942), 39.
76. Wesley Gilbertson, "Sanitary Aspects of the Control of the 1943–1945 Epidemic of Dengue Fever in Honolulu," *American Journal of Public Health and the Nation's Health* 35 (March 1945): 261.
77. Robert Usinger, "Entomological Phases of the Recent Dengue Epidemic in Honolulu," *Public Health Reports* 59, no. 13 (1944): 423–424. See also G. W. Wilson, "Epidemic of Dengue in the Territory of Hawaii during 1903," *Public Health Reports*, no. 19 (1904): 68.
78. Gilbertson, "Sanitary Aspects," 261.
79. Ibid., 262.
80. Ibid., 268, 270.
81. D. Rees, "Relationship between Military and Local Agencies Engaged in Mosquito Abatement Work in Utah," *PCMCA* 13 (1944), 121.
82. Quoted in Humphreys, *Malaria*, 143.
83. J. W. Mountin, "A Program for the Eradication of Malaria from the Continental United States," *Journal of the National Malaria Society* 3, no. 1 (1944): 69.
84. Freeborn, "The Malaria Control Program," 22.
85. Freeborn, "Military Mosquito Control," 60.
86. C. T. Williamson, "Eastern Association of Mosquito Control Workers and Its Activity during 1939," *PNJMEA* 27 (1940), 95.
87. Thomas D. Mulhern, "Draft: Chronological Historical Notes American Mosquito Control Association, Inc." (1963, 12), Thomas Mulhern's Papers, Red Binder Mosquito and Vector Control Association of California Archives, Greater Los Angeles Vector Control District, Santa Fe Springs, California.
88. Thomas J. Headlee, "House Organ of the Eastern Association of Mosquito Control Workers," *MN* 1 (1940): 2.
89. R. D. Glasgow, "Letter to Thomas Mulhern" (Eastern Association of Mosquito Control Workers, 1942), 2–3.
90. Ibid., 3, 7.
91. Ibid., 7.
92. William B. Herms, "Medical Entomology and the War," *Journal of Economic Entomology* 38, no. 1 (1945): 11.

Chapter 8 **The Postwar Era**

1. C. B. Hutchinson, "University Welcome," *PCMCA* 17 (1949), 2.
2. William B. Herms, "Looking Back Half a Century for Guidance in Planning and Conducting Mosquito Control Operations," *PCMCA* 17 (1949), 89.
3. Ibid., 93.
4. Ibid., 89.
5. H. F. Gray, "President's Address," *PCMCA* 17 (1949), 28.

6. Ibid., 27.
7. Ibid., 28.
8. Maurice Provost, "Report on the Field Trip Made through the State of California to Observe Mosquito Control Operations," *PFAMA* 20 (1949), 26.
9. Ibid.
10. Ibid., 27.
11. F. C. Bishopp, H. H. Stage, and Helen Sollers, "A Review of Mosquito Work throughout the World in 1948," *PNJMEA* 36 (1949), 16.
12. "Over 100,000 Pounds of DDT Applied during F.Y. 1946–1947," *Mosquito Buzz* 2, no. 1 (1948): 1.
13. Bennett E. Tousley, "President's Address," *PNJMEA* 33 (1946), 6.
14. G. H. Bradley, "Mosquito Control Activities of the Communicable Disease Center, U.S. Public Health Service," *PNJMEA* 34 (1947), 50.
15. Margaret Humphreys, *Malaria: Poverty, Race, and Public Health in the United States* (Baltimore: Johns Hopkins University Press, 2001), 152.
16. John A. Mulrennan, "Malaria Control," *Forty-sixth Annual Report of the State Board of Health* (Jacksonville, Florida: 1946), 49. See also Gordon Patterson, *The Mosquito Wars: A History of Mosquito Control in Florida. Gainesville: University Press of Florida, 2004*, 106–109.
17. William C. Reeves, "Perspectives on Mosquito Research by the University of California—Past, Present, and Future," *PMVCAC* 60 (1992), 4.
18. R. D. Glasgow, "The Growing Public Interest in Mosquito Control and Related Work," *PNJMEA* 34 (1947), 97.
19. H. F. Gray, "The Expanded Program of Mosquito Control in California," *PNJMEA* 36 (1949), 97.
20. Richard F. Peters, "Dick Peter's Oral History," 17. Mosquito and Vector Control Association of California Oral History Collection, Santa Fe Springs, California, 1994.
21. Ibid., 18.
22. Ibid., 19.
23. Ibid., 19–20.
24. Ibid., 20.
25. Ibid.
26. William C. Reeves, "Arbovirologist and Professor, UC Berkeley School of Public Health: An Oral History Conducted in 1990 and 1991 by Sally Smith Hughes," 223. Regional Oral History Office, Berkeley, Bancroft Library, University of California, 1993.
27. Ibid., 224.
28. "Association News," *MN* 7, no. 4 (1947): 162.
29. Earl W. Mortenson, "Earl Mortenson's Oral History," 5. Mosquito and Vector Control Association of California Oral History Collection, Santa Fe Springs, California, 1996.
30. Ibid., 15.
31. Ibid.
32. Ibid., 7.
33. Austin W. Morrill Jr., "News and Notes," *MN* 16, no. 4 (1956): 312.
34. Bailey Pepper, "Report of the Nominating Committee: Discussion," *PNJMEA* 31 (1948), 133.
35. D. J. Sutherland, "Obituary: Daniel M. Jobbins 1910–1979," *MN* 39, no. 1 (1979): 155.
36. Pepper, "Report of the Nominating Committee," 133.

37. Lewis T. Nielsen, "The Utah Mosquito Abatement Association: A Brief History of Mosquito Control in Utah," *PUMAA* 15 (1962), 15.
38. J. H. Bertholf, "DDT Resistant Mosquitoes in Broward County, Florida," *PFAMA* 21 (1950): 82.
39. Thomas Cain, "Observations on DDT-Resistant Species of Mosquitoes Found in Brevard County," *PFAMA* 21 (1950): 84. See also Patterson, *Mosquito Wars*, 113–116.
40. "Report of the Secretary," *PFAMA* 21 (1950): 5. The USDA laboratory remained in Orlando until 1963 when it was relocated to Gainesville. A number of important discoveries are associated with this laboratory. These include the use of ultralow-volume adulticiding spray techniques and the discovery of the insect repellent DEET. See Florida Coordinating Council on Mosquito Control, "Florida Mosquito Control: The State of the Mission as Defined by Mosquito Controllers, Regulators, and Environmental Managers" (Gainesville: University of Florida, 1998).
41. Maurice Provost, "The Research Program of the Florida State Board of Health's Bureau of Entomology," 2–3. Maurice Provost's Papers, Florida Medical Entomology Library, University of Florida, Vero Beach, Florida, 1954.
42. See Gordon Patterson, "Butter Knives, Watermelon Seeds, and Entomological Research," *Wing Beats* 17 (2006): 24–31; and Patterson, *Mosquito Wars*, 117–127.
43. Maurice Provost, "The Florida Medical Entomology Laboratory: How It Came to Be Where and What It Is," 5. Maurice Provost's Papers, Florida Medical Entomology Library, University of Florida, Vero Beach, Florida, 1975.
44. Maurice Provost, "The Proposed Entomological Research Center of the Florida State Health Board," 1. Maurice Provost's Papers, Florida Medical Entomology Library, University of Florida, Vero Beach, Florida, 1953.
45. Ibid., 4.
46. "News and Notes," *MN* 11, no. 1 (1951): 48.
47. Maurice Provost, "Man, Mosquitoes, and Birds," *Florida Naturalist* April (1969): 64.
48. Glenn Collett, interview with author, July 21, 2005.
49. Ibid.
50. Reeves, "Arbovirologist and Professor," 254.
51. Wilton L. Halverson, William A. Longshore, and Richard F. Peters, "The 1952 Encephalitis Outbreak in California (Preliminary Draft)" (Berkeley: California Department of Public Health, 1953), 2.
52. Ibid.
53. Reeves, "Arbovirologist and Professor," 256.
54. Halverson, Longshore, and Peters, "1952 Encephalitis Outbreak," 1.
55. Reeves, "Arbovirologist and Professor," 259.
56. This absurd formulation was intended to convey the review panel's commendation of the efforts to improve the technical expertise and competency of local agencies. Communicable Disease Center Region IX, "Report to the Director of Public Health California Department of Public Health on a Program Review Bureau of Vector Control," (Atlanta and San Francisco: CDC and Region IX, U.S. Public Health Service, May 1956), 17.
57. Ibid., 32.
58. Maurice Provost, "Birds and Mosquito Control in Florida," 2. Maurice Provost's Papers, Florida Medical Entomology Library, University of Florida, Vero Beach, Florida, 1951.
59. Florida Coordinating Council, "Florida Mosquito Control," 116.

60. CDC, "Report to the Director," 45–46.
61. Ibid., 60.
62. Jesse B. Leslie, "Behind the Scenes with the Mosquito Study Commission of the State of New Jersey," *PNJMEA* 43 (1956), 16.
63. Jesse B. Leslie, "Our Fiftieth—a Tale of Fifty Years of Anti-Mosquito Activities in New Jersey," *PNJMEA* 50 (1963), 438.
64. William Dillistin, "The Creation of the State Mosquito Control Commission," *PNJMEA* 44 (1957), 8.
65. Leslie, "Behind the Scenes," 17.
66. Quoted in ibid., 19.
67. Dillistin, "State Mosquito Control Commission," 9–10.
68. Leslie, "Behind the Scenes," 17.
69. See Daniel Cohen and Oscar Sussman, "Equine Encephalomyelitis in New Jersey," *PNJMEA* 44 (1957), 79–80; Leslie D. Beadle, "An Appraisal of the Arthropod-Borne Viral Encephalitides and Their Possible Significance to New Jersey," *PNJMEA* 47 (1960), 61; and, Paul. P. Burbutis and D. M. Jobbins, "*Culiseta Melanura* Coq. and Eastern Equine Encephalomyelitis in New Jersey," *PNJMEA* 44 (1957), 76.
70. Wayne Crans, interview with author, May 9, 2006.
71. Burbutis and Jobbins, "*Culiseta Melanura*," 68–69.
72. Jordi Casals, the Rockefeller Foundation scientist who identified alpha viruses and flaviviruses (yellow fever, dengue, etc.) coined the word "arbovirus" to describe EEE, SLE, WEE, and other insect-borne viruses in 1957. The word's origins go back to Reeves and Hammon's work in 1940. During their first summer in Yakima Valley, Reeves felt it incumbent on him to give Hammon, a physician, a brief introduction to entomology. Since mosquitoes were the only vectors that they identified in the transmission cycle, they used the term "mosquito-borne encephalitides" to describe the disease. Reeves knew, however, that other insects such as ticks and fleas also vectored diseases. Hammon and Reeves adopted the expression "arthropod-borne virus encephalitides" in the mid-1940s to describe this complex of diseases. The term continued to evolve over the next decade. Reeves felt that all the expressions were "too much of a mouthful." Reeves hit on the idea of combining the first two letters of "arthropod" with the first three letters of "borne" and joining them to the word "virus" to form "arborvirus." He liked the term. Problems emerged when he used the expression at a WHO international conference. Participants looked confused when Reeves described his latest work with "arborviruses." What, Reeves was asked, do these viruses "have to do with trees?" Jordi Casals suggested dropping the "r" creating the word "arbovirus." See Reeves, "Arbovirologist and Professor," 119.
73. William Dillistin, "Activities of the State Mosquito Control Commission," *PNJMEA* 47 (1960), 16.
74. Maurice Provost, "Florida Investigates the Life History of *Aedes Taeniorhynchus*," 2–3, 5. Maurice Provost's Papers, Florida Medical Entomology Library, University of Florida, Vero Beach, Florida, 1954.

Chapter 9 **Discontent and Resistance**

1. Stanley B. Freeborn, "The Malaria Control Program of the U.S. Public Health Service among Civilians in Extra-Military Areas," *Journal of the National Malaria Society* 3, no. 1 (1944): 23.
2. A. W. Buzicky, "Organized Mosquito Control in Minnesota," *PCMCA* 29 (1961), 87.

3. Ralph Barr, *The Mosquitoes of Minnesota* (Minneapolis: University of Minnesota Press, 1955), 5.
4. Buzicky, "Organized Mosquito Control," 88.
5. C. L. Meek, "Les Maringouins de Meche and the Legacy of Two Men," *Journal of the American Mosquito Control Association* 7, no. 3 (1991): 376.
6. C. L. Meek, "History of Mosquito Control in Louisiana," *Mosquito Control Training Manual* (New Orleans: Louisiana Mosquito Control Association, 1993), 13–14.
7. Meek, "Les Maringouins," 373.
8. Austin W. Morrill Jr., "News and Notes," *MN* 15, no. 2 (1955): 128.
9. Edward S. Hathaway, "The Louisiana Mosquito Control Association Examines Its Problems and Charts a Course," *PNJMEA* 45 (1958), 43–44.
10. Ibid., 44.
11. Meek, "Les Maringouins," 377.
12. Meek, "History of Mosquito Control," 15.
13. Charles Anderson, interview with author, November 8, 2006.
14. Thomas D. Mulhern, "Greetings from the AMCA," *PNJMEA* 55 (1968), 45.
15. Meek, "Les Maringouins," 378.
16. F. W. Harden, H. R. Hepburn, and B. J. Etheridge, "A History of Mosquitoes and Mosquito-Borne Diseases in Mississippi 1699–1965," *MN* 27, no. 1 (1967): 63.
17. F. W. Harden, "Mosquito Control at NASA's Mississippi Test Operations," *MN* 25, no. 2 (1965): 123.
18. Jack Salmela and E. A. Philen, "A Cooperative Mosquito Control Plan for Cape Canaveral and the NASA Merritt Island Launch Area Involving Federal, State, a Local Agencies," *MN* 24, no. 1 (1964): 15.
19. Ibid.
20. Ibid., 16.
21. Ibid., 17.
22. Andrew J. Rogers, interview with author, May 18, 2000.
23. The Department of Interior Fish and Wildlife citation read: "This award is the highest honor bestowed by the Secretary to private citizens and groups for direct contributions to the mission and goals of the Department. It was presented to Mr. Salmela for his endless contributions to wildlife conservation through effective mosquito control techniques and his personal dedication to effective management of wildlife resources." See Florida Coordinating Council on Mosquito Control, "Florida Mosquito Control: The State of the Mission as Defined by Mosquito Controllers, Regulators, and Environmental Managers" (Gainesville: University of Florida, 1998), 185 and Appendix I.
24. T. Wayne Miller, "Lee County Mosquito Control Report," *PFAMA* 34 (1963), 155. See also Gordon Patterson, *The Mosquito Wars: A History of Mosquito Control in Florida*. Gainesville: University Press of Florida, 2004, 128–131.
25. Maurice Provost, "Man, Mosquitoes, and Birds," *Florida Naturalist* April (1969): 64.
26. Miller, "Lee County Mosquito Control Report," 160.
27. "J. N. 'Ding' Darling National Wildlife Refuge," U.S. Fish and Wildlife Service, http://www.fws.gov/dingdarling/About/About.htm.
28. "J. N. 'Ding' Darling National Wildlife Refuge," Sunny Day Guide, http://www.sunnydayguide.com/sanibel_captiva/features/dingdarl.html.
29. Fred Lowe Soper, "Aegypti and Gambiae: Eradication of African Invaders in the Americas," *PNJMEA* 50 (1963), 158–159.
30. Ibid., 154.
31. Ibid.

32. Marston Bates and Manuel Roca-Garcia, "Laboratory Studies of the Saimiri-Haemagogus Cycle of Jungle Yellow Fever," *American Journal of Tropical Medicine* 25, no. 3 (1945): 203.
33. Fred Kiser, "Comments on the Accomplishments of Fred Soper," *MN* 37, no. 2 (1977): 307.
34. Soper, "Aegypti and Gambiae," 159.
35. Fred Lowe Soper, "More About Malaria Eradication," *MN* 18, no. 2 (1958): 54.
36. H. F Gray, "An Analysis of the Eradication Concept," *MN* 20, no. 1 (1960): 18–19.
37. Ibid., 22.
38. Elizabeth Etheridge, *Sentinel for Health: A History of the Centers for Disease Control* (Berkeley: University of California Press, 1992), 122.
39. Quoted in Soper, "Aegypti and Gambiae," 160.
40. Etheridge, *Sentinel for Health*, 122.
41. D. J. Schliessmann, "The *Aedes Aegypti* Eradication Program in the U.S.," *MN* 24, no. 2 (1964): 131.
42. Russell E. Fontaine, "The Current Status of the *Aedes Aegypti* Eradication Program in the United States and 1964 Summary of the Mosquito-Borne Encephalitides in the United States," *PCMCA* 33 (1965), 20.
43. Ibid., 21.
44. James H. Heidt, "Problems That Confront Mosquito Control Organizations in Florida," *PFAMA* 37 (1966), 11.
45. John A. Mulrennan, "Bureau of Entomology," In "Annual Report State Board of Health State of Florida, 1965" (Jacksonville: Florida State Board of Health, 1966), 36.
46. John A. Mulrennan, "Bureau of Entomology," In "Annual Report State Board of Health State of Florida, 1968" (Jacksonville: Florida State Board of Health, 1969), 65.
47. See Patterson, *Mosquito Wars*, 147–148.
48. Carol M. Anelli, Christian H. Krupke, and Renée Priya Prasad, "Professional Entomology and 44 Noisy Years since *Silent Spring*: Part 1," *American Entomologist* 52, no. 4 (2006): 224.
49. Ibid., 228–229.
50. Rachel Carson, *Silent Spring* (New York: Houghton Mifflin Company, 1962), 297.
51. Donald L. Collins, "Le Sacre Du Printemps—Sans Bruit," *MN* 22, no. 4 (1962): 406.
52. Quoted in Anelli, Krupke, and Prasad, "Professional Entomology," 232.
53. C. T. Williamson, "Report of the Law Suit Brought against the Suffolk County Mosquito Control Commission to Prevent the Use of DDT," *PNJMEA* 55 (1968), 52–53.
54. Dennis Puleston, "Document 113: Founding the Environmental Defense Fund," in *The Environmental Debate: A Documentary History*, ed. Peninah Neimark and Peter Rhoades Mott, (Westport, Conn.: Greenwood Press, 1999), 212–213.
55. Williamson, "Report of the Law Suit," 54.
56. Ibid., 56–57.
57. Clarence Cottam, "The Coordination of Mosquito Control with Wildlife Conservation," *PNJMEA* 25 (1938), 221.
58. W. R. Mitchell, "Environmental Problems in Illinois: Villages' Space-Spray Prohibitions Ruled Invalid by Circuit Court," *PUMAA* 26 (1973), 4.
59. Ibid., 5.
60. John A. Mulrennan, "Report of the Secretary-Treasurer," *PFAMA* 41 (1970), 7.

Epilogue

Epigraph: Maurice Provost, "Birds and Man," *Maurice Provost's Papers* (Vero Beach: Florida Medical Entomology Library, University of Florida, Vero Beach, Florida, 1954), January 9, 1961.

1. Grant Walton, "The Role of the Mosquito Control Coordination Committee," *PNJMEA* 60 (1973), 8–9.
2. Henry Rupp, "Afterword," *PNJMEA* 61 (1974), 201.
3. John Smith, "Report of the Entomological Department of the New Jersey Agricultural College Experiment Station for the Year 1901" (Trenton, N.J.: John Murphy Publishing Co., 1902), 587.
4. In 1996, the CMVCA was reorganized as the Mosquito and Vector Control Association of California (MVCAC).
5. John Beidler, interview with author, November 8, 2006.
6. John A. Mulrennan, "The Third Era of Mosquito Control Has Just Begun," *PFAMA* 45 (1973), 52.
7. Richard F. Frolli, "Resistance and Changing Approaches to Mosquito Control," *PCMCA* 39 (1971), 1, 2.
8. Leonard J. Fisher, "Mosquito Agency Corrupt, Witnesses Tell the S.C.I.," *Newark Star-Ledger*, December 2, 1970, 1, 21.
9. Ronald Sullivan, "$114,000 Shakedown in New Jersey Laid to Mosquito Commission," *NYT*, December 2, 1970, 1.
10. "1970 Annual Report" (Trenton, N.J.: State of New Jersey Commission of Investigation, 1971), 69–70.
11. Ibid., 84.
12. Joseph Sullivan, "$250,000 Landfill Scandal Is Charged to 11 in Hudson," *NYT*, September 15, 1972. Kenny and two of his associates were acquitted of the charges. See "John J. Kenny and 2 Others Acquitted in Park-Land Case," *NYT*, December 21, 1973, 3.
13. Howard Emerson, interview with author, March 15, 2007.
14. U.S. Environmental Protection Agency, "Larvicides for Mosquito Control," http://www.epa.gov/pesticides/health/mosquitoes/larvicides4mosquitoes.htm#microbial.
15. Douglas Carlson, Peter O'Bryan, and Jorge R. Rey, "A Review of Current Salt Marsh Management Issues in Florida," *Journal of the American Mosquito Control Association* 7, no. 1 (1991): 84–85.
16. I am indebted to Bruce Eldridge for the references to *Cx. salinarius* and contemporary entomologists' understanding of the virus-vector relationship.
17. Judy Hansen, interview with author, March 15, 2007.
18. Judy Hansen, "Presidential Address: Our Race to the Future," *Journal of the American Mosquito Control Association* 6, no. 3 (1990): 362, 365.
19. William B. Herms, "Looking Back Half a Century for Guidance in Planning and Conducting Mosquito Control Operations," *PCMCA* 17 (1949), 93.

Index

About the Author

GORDON PATTERSON is a professor of history at Florida Institute of Technology in Melbourne, Florida, and a visiting scholar at the Center for Vector Biology at Rutgers University.